PROSPECTS FOR SUSTAINABLE DEVELOPMENT IN THE CHINESE COUNTRYSIDE

Prospects for Sustainable Development in the Chinese Countryside

The political economy of Chinese ecological agriculture

RICHARD SANDERS
University College, Northampton, UK

Routledge
Taylor & Francis Group

LONDON AND NEW YORK

First published 2000 by Ashgate Publishing

Reissued 2018 by Routledge
2 Park Square, Milton Park, Abingdon, Oxon, OX14 4RN
711 Third Avenue, New York, NY 10017, USA

Routledge is an imprint of the Taylor & Francis Group, an informa business

A Library of Congress record exists under LC control number: 99071891

ISBN 13: 978-1-138-70024-6 (hbk)
ISBN 13: 978-1-138-70023-9 (pbk)
ISBN 13: 978-1-315-20486-4 (ebk)

Contents

Acknowledgements

I am grateful to a very large number of people without whose help this book could not have been written. In China I am indebted to countless numbers of officials, cadres and ordinary villagers and farmers who patiently, warmly and cheerfully answered my questions and generously put up with me in their workplaces or homes. In China, I am particularly indebted to *Professor Bian Yousheng*, of the Beijing Municipal Institute of Environmental Protection and Research, with whom I had countless interviews between 1992 and 1996, who travelled with me on three occasions to Liu Min Ying and who was a constant source of invaluable information and inspiration, *Professor Li Zhengfang*, until recently director of the Organic Food Centre, Nanjing Institute of Environmental Science, who invited me to visit his work-unit in 1993, facilitated my first visits to Xiao Zhang Zhuang and He Heng and with whom I discussed ecological agriculture between 1993 and 1995, *Xiao Xingji* and *Wen Qiuhua*, currently director and engineer respectively of the Organic Food Centre, Nanjing Institute of Environmental Sciences, who have been, between 1993 and 1997, via the spoken word, letter and e-mail, mines of information about ecological agriculture and who facilitated my visits to Qian Wei and Teng Tou in 1995 and 1997, *Professor Cheng Xu*, currently Head of the Education Department, Ministry of Agriculture, Beijing, who invited me to visit him at the then Beijing Agricultural University in 1993 and who facilitated my first visit to Dou Dian in 1995 *Gao Peng*, until very recently, engineer in the Department of Nature Conservation, National Environmental Protection Agency, Beijing, who supported my researches in ecological agriculture with good nature, enthusiasm and information from 1993 to 1997 and who facilitated my visit to Tie Xi in 1995, *Zhou Shengkun*, associate professor at the Centre for Integrated Agricultural Development at China Agricultural University, who has been an extraordinary source of inspiration concerning the political and social basis of ecological agriculture between 1995 and 1997 and accompanied me to Han Cun He in 1997, *Zhang Kuichang*, deputy Party secretary and farm manager in Liu Min Ying who cheerfully welcomed me to his village on no less than nine occasions between 1992 and 1997 and who provided me with enormous amounts of information critical to this book, *Zhang Jiashun*, deputy Party secretary of Yingshang County, Anhui Province, who welcomed me and hosted my visits to Xiao Zhang Zhuang every summer from 1993 to 1997, *Zhang Baocun*, Party secretary (1993) and *Zhang Tailin*, of the Tia Xian EPB, Jiangsu Province, who welcomed me and hosted my visits to He Heng in the summers of 1993, 1995,1996 and 1997, *Zhang Zhenliang*, Party secretary and *Zheng Deyao*, reception centre manager of Dou Dian who welcomed me and hosted my visits there in the summers of 1995, 1996 and 1997, *Fu Jialiang*, Party secretary and *Luo Zhengyao*, advisor in Teng Tou who welcomed me and hosted my visits there in the summers of 1995 and 1997, *Shen Huichun*, Party secretary and *Xu Weiguo*, President of the Board at Qian Wei, who welcomed me and hosted my visits there in the summers of 1995 and 1997, *Zhang Hongtian,* leader of Tie Xi, who welcomed me and hosted my visit there in 1995, *Sun Xiaofan, Xu Qiang, Wei Shengjin* and *Lin Tao,* colleagues and ex-students of Beijing Foreign Studies University, now friends, who generously gave of their time and skills to interpret for me whenever I asked, *Xia Qing,* ex-student of BFSU, who constantly, patiently and cheerfully worked on my behalf to set up interviews, make telephones calls, interpret, translate and effectively manage my time in China, *Li Xiaoting,* ex-student of BFSU, who was the first person to give me encouragement and contacts for my original research in China in 1992 and *Zhao Donghong,* ex-student of BFSU who accompanied me and translated for me on my first travels into the Chinese countryside in the summer of 1993, *Li Youwen,* ex-colleague of BFSU and friend who accompanied me into the Chinese countryside in 1997 and frequently interpreted and translated for me, and *Xiong Ying,* ex-student of BFSU, who travelled with me on my visits to the Chinese countryside in 1994, 1995 and 1996, and used her enormous talents as an interpreter on my behalf on many occasions.

In the UK, I am particularly indebted to *Professor John Taylor,* head of the Centre for Chinese Studies at South Bank University who has been a constant source of encouragement and good advice, *Gerry Foley,* energy consultant, author and friend, who gave me a great deal of information concerning rural energy in general and biogas in particular and who read and gave me helpful comments on chapter 2, *Professor Peter Nolan,* of Jesus College, Cambridge and ex-university classmate, who provided considerable encouragement and useful contacts in China and who read and gave me helpful comments on chapters 3 and 4, *Professor Chris Jenks,* Pro-Warden of Goldsmith's College, London University, who as a friend gave me considerable encouragement and who read and gave me helpful comments on chapter 7, *Professor Joshua Bamfield,* until recently Head of the School of Business and *Diane Hayes,* Dean of the Faculty of Management and Business at University College, Northampton, who between 1994 and 1997 always gave me time, money and support for my researches for this thesis, *The British Council,* who provided me with a senior advanced scholarship in conjunction with the *Chinese State Education Commission* to do research in China from April to September 1995, and last, but not least, my wife, *Sue,* who badgered me to do research from the start, who accompanied me to China in 1991-2, put up with my absence from the UK throughout 1992-3 and during recent summers when I have been engaged in research and who has been throughout this period a source of love and support.

1 The Problems in Context

1.1 The Reforms of Deng Xiaoping

The Peoples Republic of China is presently undergoing fundamental change. Since the historic decisions made by the 3rd Plenum of the 11th Central Committee of the Chinese Communist Party in December 1978 *(sheyijie sanzhong quanhui)*, decisions clearly authorized and pushed through by Deng Xiaoping and his supporters on the Party's "liberal" wing (Harding, 1987, p.90), China has been implementing a programme of reform and opening to the outside world *(gaige kaifang)* which has had far-reaching effects on the nature of social organisation in the countryside, has significantly expanded the role of markets in all spheres of economic activity and converted China from being a country largely isolated from the global economy to one which, if measured by the proportion of merchandise trade to GNP, has one of the more open economies in the world, ahead of Japan, USA and Germany (World Bank, 1996, pp.212-3). These reforms have been accompanied by such rapid rises in industrial output, foreign trade, GNP and GNP per head that Deng's optimistic target, announced at the 12th Congress of the Communist Party in China in 1980, of quadrupling real GNP by the year 2000 (Liu Guo Guang et al., 1987, p.28) has already been achieved comfortably ahead of time (China Statistical Yearbook, 1995, p.33).

Despite the stresses and strains imposed on any society as a result of such fast rates of GNP growth, the reform-and-open policy has enjoyed almost universally popular appeal, not least because of the material advances that it has delivered to the vast majority of the Chinese people, evident in the increased availability of foods and variety of diet, improved standards of housing (particularly in the countryside) and the expanded ownership of consumer durables over the last two decades (China Statistical Yearbook, 1998, p.324). As a result of this popular appeal, and despite the active power-broking currently taking place amongst China's top leaders, it is unlikely that the reforms will be unravelled in the foreseeable future, whatever the mixed fortunes wrought on the body politic of the Peoples Republic by the recent death of Deng Xiaoping.

1.2 Extant Political-economic Problems

Although there has been, and remains, wide popular support for the general thrust of the reform-and open-policy, the years since its inception have not been without political conflict and economic difficulty. The incidents in Tiananmen Square of Spring 1989 which came to a head on June 4 of that year bear witness to a level of political turmoil which, some observers believe, resulted directly from the very process of reform itself (e.g. Hinton, 1990, p.188, Muldavin, 1992, p.24). Whatever the precise nature and causes of those well-publicised events, (and despite the hard-line reactions of the government to them) several years later there still exist serious economic problems which cause alarm and which are responsible for simmering discontent amongst sections of the Chinese people which periodically boils over in demonstrations of one kind or another (Chen, 1997, p.107). These problems are diverse and include rapid inflation, burgeoning unemployment and widening inequalities between regions, between provinces, between the countryside and the town and between individuals, inequalities which condemn a substantial minority of the population, particularly in the countryside, and despite a general and significant rise in living standards, to a life of abject poverty. (World Bank, 1992c, p.1, World Bank 1995, p.1, Rozelle, 1996, p.64-5)

1.3 Environmental Problems

Despite the general rise in living standards, a further problem is manifested which threatens their sustainability and the quality of life engendered for the majority: the problem of environmental pollution and degradation. While few observers would deny that the problems of inflation, unemployment and inequality, if not created by the reform process begun in 1978, certainly became more prominent subsequently, environmental degradation clearly predates it.

The Chinese government, even in the early 1970s, was sufficiently alive to the potentially damaging implications of economic growth on the natural environment that it took an active part in the first United Nations Conference on Environment and Development held in Stockholm in 1972, held the first national conference on environmental protection in Beijing in 1973, established the Leading Group on Environmental Protection under the authority of the State Council in 1974, formulated its own 'Basic Law of Environmental Protection' in 1979 and established the new Ministry of Urban and Rural Construction and Environmental Protection in 1982. Environmental Protection was even written into the Constitution (CCP, 1987). Yet despite this apparent official concern and activity, it is evident that, in hindsight, the economic growth experienced both before and after 1980 was achieved at the expense of an alarming and increasing cost to the natural environment. When Vaclav Smil (1984) catalogued China's woeful litany of environmental problems, deforestation, soil erosion, salinization, desertification, and water and atmospheric pollution to name but a few, (and indicting western apologists as well as the Chinese leadership for their part in producing and reinforcing them) it became clear, both within China and without, that a sizeable environmental price was being paid for its material advance.

The early reforms, which included the sweeping away of the communes, the de facto privatization (Hinton, 1990, p.9) of the land with the reintroduction of family farming through the household contract responsibility system (*jiating chengbao ziren zhi*), the increased tolerance of rural side-line industrial enterprises, increased official prices for grain and an expanded role for rural free markets were all designed to impact first and most powerfully in the countryside. And they did so with dramatic success to be seen in increased agricultural productivity and an explosive growth in the output of side-line activities, including animal husbandry and industrial enterprises to such an extent that the increased output of rural China has been the backbone of China's 222 per cent GNP growth since 1978. (Per Ronnas, 1994, pp.232-8)

But that growth has been achieved at the expense of an almost universal blight on the natural environment. Not only has the mushrooming of rural industrial enterprises had

2

predictable implications for the intensity of noise, water and atmospheric pollution, but deforestation has led to problems including soil erosion and changes in local micro-climates with alarming effects on water-tables, causing, according to locals, floods in some areas and drought in others, while the increased use of chemical fertilisers and pesticides on crops has increased the energy intensity of agricultural production (in a country critically short of energy) while threatening the long-run quality of the soil and the health of those who work on it and who consume from it. And the increase in rewards to those engaged in non-agricultural activities in the countryside, relative to rewards from the land, has led to concern for the future health of agriculture, as farmers desert the land in increasing numbers, magnifying the long-run threat to the quality of the soil and the state of the natural environment.

1.4 Rural Sustainability

Thus while the countryside has seen rapid increases in production of all kinds, leading to major advances in material standards of living for the majority since the reforms of the late 1970s, problems have been thrown up which threaten the *sustainability* of those advances including, on the one hand, the increased costs of maintaining soil quality and productivity, particularly for grain cultivation, and on the other, some neglect of the land resulting from the migration of villagers from the field to the factory (whether in the village itself or in the local township). These problems impact upon each other and lead to a questioning of the ability of the rural economy to sustain for future generations the material developments enjoyed by the present one.

1.5 Chinese Ecological Agriculture

One of the early responses to the environmental and other problems of Chinese agriculture by the authorities was the encouragement of Chinese Ecological Agriculture (CEA) *(Zhongguode shengtai nongye)* in the early 1980s. The practices involved were hardly new, borrowing heavily from traditional organic techniques practised by Chinese peasants down the centuries. But they represented a systematic and comprehensive response to many of the problems experienced in the Chinese countryside at the time. CEA, often encouraged by research institutes and agencies was adopted in an increasingly large number of 'model' villages through the 1980s and the experiment was given a shot in the arm when the very first of these model villages, Liu Min Ying, in Daxing County, in the SE suburbs of Beijing, was awarded a United Nations Environmental Programme (UNEP) global honour for environmentally friendly economic development in 1987.

Much has been written about CEA by Chinese scholars and specialists in the field (e.g. Fang 1995, Tao 1993) but often in a propagandist style, and based upon the unspoken assumption that ecological agriculture is 'a good thing' and thus generally worth propagandizing for and that its extension is largely unproblematic, given sufficient education of farmers as to its benefits. The literature contains little close observation and analysis of the implications for the economy and environment of those villages who have adopted it in the past and still less on the political-economic, social conditions necessary for its *successful* adoption, implementation and continuation and thus on the likelihood of its extension to other areas of the Chinese countryside.

1.6 Aims of the Research

The aims of the research for this book have been to correct these deficiencies in the literature, specifically, to answer three interlinked questions:

(1) Is Chinese Ecological Agriculture worth adopting? Specifically, does CEA promise a form of *sustainable* rural development?

(2) To the extent that it does, what are the social, political and economic conditions in the Chinese countryside which most favour its extension?

(3) To the extent that these conditions are restrictive, what can the Chinese authorities do to make them less so and thus encourage its extension?

It is self-evidently the case that, however proficient a new productive process in the countryside is likely to be, it is of no practical significance if farmers won't adopt it for one reason or another. Not only should new agricultural processes and initiatives be *technically* proficient and attractive but they should also be *socially* acceptable to farmers. As Zhou Shengkun of the Centre for Integrated Agricultural Development at the China Agricultural University said of some of the research done in China today in justification of his institution's socio-economic perspectives on agricultural development (1995, private conversation):

> There are too many people engaged in agricultural research of a scientific nature and coming up with new techniques, new strains and so on which are technically feasible but without thinking enough about whether the farmers will actually adopt them.

This book investigates the experiences of seven villages and two counties in very different parts of the Chinese countryside which have adopted CEA in the last two decades. In particular, it examines the success of those units in doing so in the light of the social, political and economic factors that define the parameters within which farmers in the countryside actually live their lives and which determine the degree of acceptability that a new government-sponsored initiative like CEA will have. To the extent that the changes wrought by the rural reforms, in particular the almost universal sweeping-away of the communes and the increased role for markets, have had a major influence over those parameters, this book examines the adoption of ecological agriculture with particular reference to those reforms and in the light of the responses of the villagers to the reintroduction of family farming and the implementation of the HRS.

4

1.7 Plan of this Book

Chapter 2 introduces the tricky and often elusive concepts of growth, development and sustainability and, in the light of a literature review, defines rural sustainable development as *development in the countryside which threatens neither life nor livelihood in the medium or long term.*

Chapter 3 begins by reviewing the rural political economy of China since 1949, both in terms of social and political forms and of economic growth and development, concluding with a full discussion of the rural reforms in the late 1970s and early 1980s and their impact on the rural political economy.

Chapter 4 examines the implications for sustainability of the rural political economy consequent upon the rural reforms of the late 1970s and early 1980s. It concludes that there are serious problems threatening that sustainability which stem at least partly from the reform process itself.

Chapter 5, with the aid of a comprehensive literature review, discusses the state of the Chinese environment in general and the steps taken by the Chinese government with regard to environmental protection to date. Chapter 6 continues by discussing Chinese Ecological Agriculture as one response by the authorities to environmental and other problems perceived to be threatening the health of the Chinese rural economy. It includes a review of the literature on CEA in English translation and describes its introduction in the early 1980s and its implementation and extension in the Chinese countryside to date.

Chapter 7 discusses the design of the research for this book and reviews the debates around the appropriate methodology for 'doing' social science research. It explains and defends the qualitative, interpretative and anthropological approaches adopted as a means of researching the implications for Chinese villages and counties of introducing CEA, approaches which rely heavily on case studies, observation and interviewing. The concept of the researcher as 'bricoleur' (see Levi-Strauss, 1962, p.17) is explained and defended. Also defended is the approach of political economy, defined as one which takes as axiomatic that political forms and structures are both determined and determining, determined by economic structures and determining economic and social outcomes. The chapter will conclude by discussing the particular problems thrown up by doing social research in the PR China.

Chapter 8 presents evidence, drawn from original fieldwork, of the implications for seven villages (and, to a lesser extent, two counties) across the Chinese countryside of adopting CEA. It discusses the developments made, the gains won, the difficulties incurred and it assesses the benefits of ecological agriculture to the villagers involved. Additionally, it discusses the political economy of each village since 1949, investigating the history of recent economic, political and social change and in particular, discusses the responses of each village to the rural reforms of the late 1970s and early 1980s.

Chapter 9, on the basis of the evidence of the previous chapter, concludes with answers to the questions posed in 1.6 above. Specifically, it argues:

(1) That CEA, despite certain difficulties and problems, holds out the prospect of a more

sustainable future for the rural economy than more usual forms of activity in the Chinese countryside.

(2) That the conditions under which CEA is likely to be adopted in the Chinese countryside are restrictive and that one of the most serious handicaps to the possibility of its extension is the spread of small-scale, privatized family farming since the rural reforms.

(3) That if the Chinese government are serious about the desire to extend CEA in the Chinese countryside, the authorities must face up to the need to address the land question anew. While an authoritarian return to the commune system is unthinkable, the possibility of new forms of land management and ownership, putting greater stress on the importance of larger scale production units, on collective rather than household norms and on longer rather than short-term horizons should be urgently investigated.

2 Growth, Development and Sustainability

2.1 Overview of the Research Problem

One problem in producing work in the context of environment and development is the difficulty of terminology. This is because, in common with other terms frequently used in the debate surrounding them and in economics and policy-making generally, *environment* and *development* are highly charged, contentious concepts, the meaning of which can only ever be understood partially, in both senses of that word. In that such meanings are socially constructed and negotiated, they allow a space wherein, more often than not, powerful groups are able to impose meaning. This work attempts an engagement in that latter process.

2.2 Economic Growth

Economic growth is generally accepted to mean increases in real gross national product (GNP) per capita. Notwithstanding the immense difficulties (and associated controversies) involved in measurement, particularly in low income countries, growth in GNP per head is generally regarded as an acceptable representation of increases in the size of the national cake, or command over material goods and services, that a country's inhabitants can make. This is not to say that GNP is necessarily a good measure even of the (dollar) value of production in that the statistic can only include what it is possible to measure. Goods and services which are bought and sold in the market-place are thus normally included in this category while a vast quantity of goods and services outside of the cash nexus, including those produced in the home and/or exchanged informally are not and therefore excluded from the statistics. Omissions of this kind are likely to be significant in those countries, often poorer ones like China, where the size of the informal sector sector is disproportionately high. Indeed, this particular problem has led to heated debates as to the very size of the Chinese economy today (World Bank, 1995, p.3).

2.3 Gross National Product and Welfare (or Standards of Living)

In post-war years, it has become commonplace to use growth of GNP per head as a surrogate measure for increases in *welfare* or *standards of living* in a country while comparisons between the levels of GNP per head have been used to compare the welfare between one country and another. Welfare is important to conceptualize in that increases in it would be universally judged as desirable by politicians and seen by economists as the main objective of their theorizing. But welfare, unlike material output, is a qualitative term and is understood by most observers as involving a range of tangible and intangible factors, including command over goods and

7

services in a variety of forms, both public and private, and opportunities for personal and communal fulfilment and emancipation. Being an essentially qualitative term, it does not easily or satisfactorily lend itself to quantitative measurement. But this has not dampened the enthusiasm of many observers, particularly of economists, for doing so.

Until the 1960s, with the publication of mainstream literature which for the first time seriously questioned the growth process (e.g. Mishan,1967), few observers, looked past increases in GNP for that measure. Indeed, early development economics was, to a greater or lesser extent, predicated on the assumption that growth *meant* development, that economic growth and development were essentially the same thing. Even Andre Gunther Frank, who questioned a great deal of development orthodoxy in the 1960s from a radical point of view (Frank, 1966) admitted by the 1990's (Frank,1992, p.136), that he had:

> maintained the essence of the thesis that economic growth through capital accumulation equals development.

By 1962,the UN, clearly concerned with the difficulties in defining and measuring development considered that:

> development is growth plus change. Change, in turn, is social and cultural as well as economic, and qualitative as well as quantitative. The key concept must be improved quality of peoples' life. (Esteva, 1992, p.13)

More recently, in 1990, the UNDP's first Human Development report was published in which 'human development' as a process was defined as 'the enlargement of relevant human choices' and an acceptable level of achievement as 'the internationally compared extent to which, in given societies, those relevant choices are actually attained.' (Esteva, 1992, p.17). The nowadays annual UNDP Human Development Report produces a Human Development Index which combines three measures, life expectancy, adult literacy and real GNP per capita in a single number.

But while there is nowadays a broad measure of agreement that there are problems in the use of GNP statistics as measures of welfare and that attention must be paid to other aspects of the quality of life, Latouche (1992, p.253) argues,

> the practical import of such wider conceptions has been largely symbolic. Even where they have led to concrete action in favour of basic needs, self-sufficiency in food production or appropriate technologies, their overall impact has been negligible. The results have not been without ambiguities and have certainly not attained a sufficient salience to modify the dominant GNP perspective.

Thus economic growth conventionally measured by per capita GNP remains the most commonly accepted representation of increases in standards of living and the most important measure in any discussion of development. This is not merely the case in rich countries and in the corridors of power at the World Bank: indeed the GNP statistic holds perhaps even more sway over the minds of the politicians in poor countries like China. Thus while this leap from growth to development via GNP involves a considerable exercise of the imagination, and even while such a leap is commonly treated with scepticism, there is still very little challenge to GNP's position in the imposition of meaning. Indeed, to keep harping on the limitations of GNP to reflect welfare would, for Arthur Koestler (1967, pp.349-353), rank alongside behavioural psychology as a dead-horse the flogging of which amounts to cruelty.

The absurdity of this state of affairs is well expressed by Chesneaux (quoted by Latouche, 1992, p.257):

> Flabbergasted, the fisherman in Samoa, who lives quite at ease in relative self-sufficiency, learns that in terms of GNP, he is one of the poorest inhabitants of the planet, while in terms of GNP, the unemployed worker in the slums of Caracas discovers with amazement that he enjoys a standard of living which is worthy of envy. (1989, p.64)

The arguments why conventionally measured economic growth is not universally perceived to be an acceptable measure of the development of welfare are well rehearsed. Not only does GNP omit large amounts of real- if unmarketed - output, but the bluntness of the GNP statistic is insensitive to the forms of output it encompasses: increases in the output of primary education for girls is lumped together with the production of candyfloss as though their implications for welfare were the same. The *distribution* of income is hidden: where it is highly skewed a mean figure may give no indication of the material welfare of the majority of the inhabitants. And GNP statistics, by ignoring inequality ignore the fact that an extra pounds worth of income to a poor household implies significantly more additional welfare than an extra pound to a rich one.

2.4 The Costs of Growth

Moreover, faced with evidence of pollution, resource depletion and environmental degradation at every turn even the most hardened growth enthusiast is not unaware that growth has opportunity costs, ignored by crude GNP statistics. The absurdity of ignoring the environmental costs of economic growth is magnified when activities which involve immediate and direct cost in terms of degradation of the environment, such as deforestation, should they involve monetary gain rewarded in the market-place, *add* to GNP rather than detract from it. As Daly (1980, p.26) suggests, this should be labelled 'swelling', not growth. There are attempts presently being made in the UK (Pierce et al., 1989, pp.117-8), to refine the calculation of GNP to include environmental costs, in the same way that the Net National Product figure in the national income accounts tries to take into account the depreciation of fixed capital. Goldemberg discusses methods of constructing a figure for 'Green GDP' (1996, p.126). There have, over the years, been several attempts to devise appropriate accounting systems for developing countries, most notably by the United Nations Statistical Office (World Bank, 1992a, p.35), but a successful conclusion to these attempts appears a long way off and, in the meantime, it is instructive to note that the conventional NNP figure, while being the bottom line in the conventional national income accounts, has little or no significance in comparison to Gross National Product when discussing issues pertaining to growth and development.

2.5 Development

It is clear that the leap of imagination from growth to development must at the very least take place within the framework of such questions as 'growth of what?', 'growth to whom?', and 'at what cost?' (Donaldson, 1973, ch. 6) and this latter question allows the debate about the very meaning of the term 'development' to be opened up. The project of Sachs et al. (1992) is no less than the very deconstruction of the concept of development itself. For Sachs, the notion of development and its underbelly, underdevelopment, were invented on January 20, 1949, by President Harry S. Truman, when he announced:

> We must embark on a bold, new program for making the benefits of our scientific advances and industrial progress available for the improvement and growth of the underdeveloped areas (Esteva, 1992, p.6).

Gestava Esteva (1992, p.7) explains:

> Underdevelopment began then, on January 20, 1949. On that day 2 billion people became underdeveloped. In a real sense, from that time on, they ceased being what they were, in all their diversity, and were transmogrified into an inverted mirror of others' reality: a mirror that belittles them and sends them off to the end of the queue.

Representations of development, Esteva argues, are coloured by meanings, sometimes unwanted, based on the assumption that development always implies a move from a less favourable or inferior state to a more favourable or superior one. If a country is developing it is 'doing well, because (it) is advancing in the sense of a necessary, ineluctable, universal law' (1992, p.10).

If the immediate post war years saw a concerted attempt by the United States of America and international forums and agencies such as the United Nations and World Bank (officially titled the International Bank for Reconstruction and *Development)* to launch an era of development, it has manifestly failed across many parts of the globe. Fifty years on, gaps between rich and poor countries have widened, with many countries absolutely poorer in any representation of that term. Using the World Banks's basic indicators (and mindful of their manifest limitations) of the 30 'poorest' countries (21 in Africa) with measured GNP per capita varying from 80 to 420 US dollars per annum, no less than 13 countries recorded falls in that figure between 1965 and 1990. In 16 of these countries in 1990, life expectancy at birth remained less than 50 and female literacy in the same year was more than 50% in only four countries and less than 75% in a majority of them (World Bank, 1992a, pp.218-9). That some countries, like the East Asian 'tiger economies', have moved up the world league tables of GNP statistics over the years is not in doubt: GNP statistics have allowed some degree of mobility in inter-country ranking. What is clear from the published data, however, is that the era of 'development' has been a chimera for many people.

Even when greater material rewards have been forthcoming, they have often been achieved at great social cost. It is this realisation which has, at best, made the concept of 'development' problematic and, at worst, brought it into disrepute. Indeed, the concept of economic development carries so many negative charges that it has led many observers not merely to

10

demand the 'dethronement of GNP' but to question the value of economic development per se (Sachs, 1993, p.5).

This book does not accept that position. It does accept, however, that the process of development should involve increases in *welfare,* notwithstanding the difficulties of definition and elucidation, and thus embraces the 'welfare critique' (Sutcliffe, 1995, p.240) of conventional development theory. It recognises that growth of GNP per head may lead to increases in welfare but equally, it may not.

2.6 Development and the Environment

In the post-war period, all countries have expressly attempted to develop their economies. Some, like China until the 1980s, have tried to do so largely in isolation from the international economic system, others have enthusiastically embraced it. In either case, what development has taken place, if any, has frequently taken place at considerable social cost, manifested most widely and intensively in costs imposed on the environment. While the concept of *development* is a problematic 'representation', the same is true of the *environment.* As Redclift (1987, p.3) argues, the environment is 'socially constructed', with meanings different across time, place and class. For those living in abject material poverty, the term can have little substantive meaning at all and is unlikely to be understood as anything conceptually separate from the condition of material poverty itself. This is hardly surprising given the fact that that those suffering most from material poverty are, arguably, those living in the most degraded of environmental conditions.

Whatever its particular representation, however, it has become increasingly clear even in rich countries that *environmental* issues are not, in and of themselves, distinct and separate from *economic* issues. Indeed, the increasing concern for the environment amongst powerful elements in western capitalist economies is illustrative of an understanding of the interlinkage between the environment and the economy, how *environmental* damage becomes *economic* damage. Indeed, this realisation with regard to poor countries is one of the primary causes of the supposed 'greening' of the World Bank in the late 1980s (Adams,1990, p.193).

2.7 The Materials Balance Approach to Economic Development

The materials balance approach to environmental economics is instructive, partly because it makes clear the untenability of treating economic and environmental concerns as discrete issues. This approach begins with lessons drawn from the first and second laws of thermodynamics. The first law explains that matter and energy are neither created or destroyed in the productive process, merely transformed from one state to another. Hence materials may, as a result of the productive process become part of the non-productive output of the system and be returned to the environment as waste. It is in this regard that the second law is relevant: as matter is transformed it moves from a state of low entropy (high usefulness) to high entropy (low usefulness). Fossil fuels, for example, become transformed in the production process into particulate emissions and exhaust gases. As Kerry Turner (1991, p.4) suggests,

11

In lay terms, entropy is a certain property of systems which increases in an irreversible process. When entropy increases, the energy in the system becomes less available to do 'useful work'. No matter recycling processes can be 100% efficient.

The irreversibility and the incontrovertibility of the second law of thermodynamics is aptly summed up by Ehrlich and Ehrlich as follows,

If (the law) did not hold the world would be truly interesting- ice cubes would be as likely to appear spontaneously in a martini as to melt in it ..and squashed cats could reassemble themselves on the highway and trot off (Daly,1980,p.40).

Once the second law of thermodynamics is recognized and the materials balance approach is adopted, according to Turner (1991,p.4), 'it is easy to see that the way humans manage their economies impacts on the environment and, in a reverse direction, environmental quality impacts on the efficient working of the economy.'

For economists, the *environment* provides four important services: (1) Most basically, the natural environment permits the very existence of life on earth, it is life- and livelihood sustaining (2) It has amenity value: it provides opportunities for the enjoyment of beauty, relaxation and recreation. (3) It contributes to the economy by providing the productive process with important inputs in the form of exploitable resources. And (4) it provides the dumping ground for the waste-products of the production and consumption processes. The problem is that economic development, *understood as economic growth,* threatens the first two services by demanding more of the environment in terms of the latter two. Once the form or pace of economic development renders the environment incapable of performing the second of the above services, the provision of amenity value, the benefits of that development must be called into question. Once economic development threatens the first of the above, its life and livelihood sustaining services, that development becomes literally *unsustainable.*

2.8 Sustainable Development

Nowadays it is commonplace to accept that past growth rates of GNP in both developed and developing countries have had negative impacts on the environment by demanding more of the earth's biophysical throughput (Ekins and Jacobs, 1995, p.22). Early concerns about the environmental impacts of growth concentrated on resource depletion (e.g. Meadows et al., 1972) but with the development of new energy sources coupled with technological changes aimed at conservation, concerns have more recently shifted towards the polluting and resource degrading implications of growth and its implications for loss of biodiversity. Today, evidence of pollution, degradation and loss of biodiversity abound at both the regional level (e.g. toxic pollution, acidification, deforestation, desertification, salinization and soil erosion, water pollution and depletion) and the global level: (e.g. the 'greenhouse effect', ozone depletion, species extinction). Ekins and Jacobs (1995, pp.9-22) provide a concise review of the symptoms of environmental unsustainability and the linkages between them. While there is still

12

considerable debate as to the extent of the problems, for example, concerning the existence and implications of the 'greenhouse effect' (Foley, 1991a, pp.19-36, Pearce 1991,p.13), there is increasing agreement that carrying on as we have done so far - 'business-as-usual' (Foley, 1991a, p.23) - with minor, if any, modifications to the growth process, will sooner or later lead to environmental catastrophe (Ekins and Jacobs, 1995, p.27). There are, of course, a number of contrasting paradigms within which these problems are cast and interpreted, but no one disputes that environmental costs are negative externalities which must be dealt with in one way of another. That is because the unsustainability of the natural environment leads eventually to the unsustainability of economic development as the life and livelihood providing services of the environment are threatened.

2.9 Contrasting Paradigms of Sustainable Development

The problem of understanding sustainable development is particularly acute because the term is employed by scholars across a wide variety of theoretical paradigms (Redclift,1987, p.37) and is used as a *representation* of so many different developmental processes as to be in danger of becoming a 'meaningless cliche' (Lele, 1991, p.612) and falling completely into disrepute as an analytical tool. Pezzey (1992, pp.55-59) presents twenty seven different definitions. And as Pearce et al. suggest (1989, p.1), 'it is difficult to be against "sustainable development". It sounds like something we should all approve of, like motherhood and apple pie.' O'Connor (1994, p.152), from a very different ideological perspective, agrees:

> Ambiguity runs through all the most important discourses on the economy and the environment today...Precisely this obscurity leads so many people so much of the time to talk and write about 'sustainability': the word can be used to mean almost anything...which is part of its appeal.

Lele (1991, p.613) adds,

> ..In short SD is a meta-fix that will unite everybody from the profit-minded industrialist and risk-minimizing subsistence farmer to the equity-seeking social worker, the pollution-concerned or wildlife-loving First Worlder, the growth-maximizing policy maker, the goal-orientated bureaucrat and therefore the vote-counting politician.

As a recent illustration of the 'profit-minded industrialist's' interest in sustainable development, Hart (1997, p.68) claims 'the achievement of sustainability will mean billions of dollars in products, services and technologies that barely exist today..... Increasingly, companies will be selling solutions to the world's environmental problems.'

While the concept of sustainable development first entered common discourse in the early 1980s as a result of the publication of the World Conservation Strategy by the International Union for the Conservation of Nature (IUCN) in 1980 (Lele,1991, p.610) it got its greatest boost with the report of the World Commission on Environment and Development (WCED) in 1987 which states,

13

Humanity has the ability to make development sustainable - to ensure that it meets the needs of the present without compromising the ability of future generations to meet their needs. (p.8)

and which continues,

Sustainable development is not a fixed state of harmony but rather a process of change in which the exploitation of resources, the direction of investments, the orientation of technological development and institutional change are made consistent with future as well as present needs. (p.9)

As such, the Brundlandt Commission regarded sustainable development as a policy objective rather than a methodology (Redclift, 1990, p.4), and called upon 'all the nations of the world to integrate sustainable development into their principles'(WCED,1987, p.363). Nonetheless, it pointed to a number of principles with which to guide states' policy actions, the first of which was the *revival of growth,* arguing that 'economic growth must be stimulated, particularly in developing countries, while enhancing the resource base (p.364).'

Thus the Brundlandt Commission took a position which challenged the 'zero-growth' orthodoxy of much of the Green movement at the time (Jacobs, 1991, p.53) and as such was consistent with the position later taken by Pearce et al. (1989) when they stated that,

the three concepts of environment, futurity and equity are integrated in sustainable development through a general underlying theme. This theme is that *future generations should be compensated for reductions in the endowments of resources brought about by the actions of present generations*........Sustainable development is about making people better off (pp.2-3).

Though Pearce et al. have come in for criticism for their paradigm of sustainable development (Pearce et al., 1991, pp.1-10, Redclift, 1994, p.5), it is largely on the grounds of its chosen methodology for valuing the environment within the context of a traditional neo-classical problematic (Pearce et al., 1989, ch.7) rather than for its underlying policy aims. Nonetheless, for its critics:

it is clear from a variety of different ideological positions, that finding technical solutions to environmental problems, including ways of costing environmental losses such as those advocated by Pearce *et al.* (1989) is ultimately self-defeating. (Redclift, 1994, p.5)

Of course, one of the reasons why the literature on sustainable development has now reached industrial proportions is because it has attracted scholars not only from a large number of different disciplines but also from many ideological perspectives. In that Marx's writings found it 'impossible to conceive of "natural" limits to the material productive forces of society' (Redclift, 1987, p.8), it is not surprising that orthodox Marxist theorizing has had difficulty in incorporating sustainable development into its model of historical materialism. Booth (1994, pp.5-6) makes clear that Marxist approaches to developmental sociology in general have reached

14

an impasse either because they have led to grand simplifications which are empirically 'wrong' or generalizations of little relevance. The underlying reasons are two-fold. On the one hand, Marxism suffers from 'theoretical disorientation', in that there has developed no middle ground between the 'polarized paradigms' of the 1970s, with basic Marxist concepts such as 'mode of production' proving incapable of consistent application, reductionist, and frequently unrelated to current policy issues. On the other, 'preoccupations, blind-spots and contradictions' of Marxist development theory can be explained by the:

> metatheoretical commitment to demonstrating that the structures and processes of less developed societies are not only explicable but also *necessary* under capitalism (emphasis added). This general formula .cover(s) two forms of 'necessitist' commitment in Marxism: the notion that the salient features of capitalist national economies and social formations can be ..'read off' from the concept of the capitalist mode of production and its 'laws of motion' and the various forms of Marxist system teleology or functionalism (1994, p.5).

Redclift (1987, p.46) argues that Marxist approaches have illuminated the debate by stressing the ways in which nature is transformed by commodity production *within capitalism,* the importance of *distributive* effects of environmental change and the *ideological* content of environmental ideas and their links to legitimation processes within capitalism. However he convincingly argues that orthodox Marxism, by presenting environmental problems as but one more illustration of the contradictions of capitalism leading to its ultimate (and welcome) demise is, at the end of the day, vacuous. As Redclift (1987, p.48) suggests, 'the costs of environmental degradation, especially in the South, are such that the final scenario of capitalism destroying itself through ecological attrition is unacceptable' while going on to say that 'the point at which the costs in destroying the environmental and non-market social relationships exceed the benefits of further commodity production has already arrived.'

Perhaps because of the ultimate futility of the crude Marxist perspective, post-structural theorists in a Marxian tradition (such as Blaikie and Brookfield, 1987, Booth 1994 and Muldavin 1996a&b) have more recently developed a paradigm of political ecology, a paradigm with which this book feels comfortable (see Chapter 7 below), which:

> focus(es) on long-term environmental effects, forces a reevaluation of the impacts of short-term coping strategies under shifting regimes of accumulation...emphasises the importance of political economy in the understanding of environmental degradation - it is an historically informed attempt to understand the role of the state, the social relations within which land users are entwined and resulting environmental changes (Muldavin, 1996a, p.237, see also 1996b, p.291).

At the other end of the ideological spectrum from crude Marxism is the approach of neo-classical economics. There are some scholars within that fold (e.g. Bernstam,1991) who still hold to the view that the environment is a commodity just like every other and that optimum outcomes for the environment (as for everything else) will be forthcoming in free markets. The dominant position within the neo-classical paradigm, however, epitomised by Pearce et al. (1989, 1992, 1993), rests on the view that markets do, in fact, *fail* when dealing with *public goods* such as the environment and that special techniques and methodologies are needed to handle it. However once these techniques have been perfected, appropriate modifications of the market can be made through government policy to allow a 'business-as-usual' approach to economic growth to continue unabated. An early defender of economic growth from this perspective was Beckerman (1974) for whom 'the environmental problem and the growth problem are (merely) special cases of the general problem of resource allocation' (1974, p.55).

For many neo-classical economists, economic efficiency occurs when the present value of net benefits of resource use are maximised and its necessary condition in the case of exhaustible resources, referred to as the Hotelling Rule (see Hotelling, 1931), is the requirement that the rate of change of resource royalty or price net of extraction cost is equal to the market rate of interest. And when *renewable* resources are involved, according to Pan (1992, p.2),

> The fundamental principle in the economics of renewable resources requires that, along a stationary policy, marginal net social benefit from increasing current harvest be equal to marginal harvest cost discounted at the net social rate of discount, which is the difference between the interest and the marginal productivity of the resource stock.

In other words, where renewable resources are used in the production process, as with any economic process, economic efficiency will be attained when the net marginal social benefits of doing so are just equal to the net marginal social costs.

But as Pearce recognises, much of the criticism of this paradigm rests on widespread scepticism, even within orthodox economics itself, of the ability of scholars to find appropriate methodologies to model the above values meaningfully. Pezzey analyses sustainability within a neo-classical paradigm and while accepting that a simple model of unsustainability can be constructed using them and that in theory, its application is straightforward at the 'system' level, he argues:

> Making sustainability operational at the project level is much harder, even conceptually. System sustainability cannot be disaggregated into project rules in the simple way that system optimality can be disaggregated into cost-benefit analysis rules for project appraisal (1992, pp.ix-x).

For many scholars there are other more fundamental objections to this approach going to the roots of the assumptions and epistemology of neo-classical economics. For Seabrook (1994, p.8), not only is (neo-classical) economics pure ideology, but is 'the agent of colonization, not only in the poor countries but in the West itself' (p.15). For Redclift (1987):

> Neo-classical economics assumes that resources are divisible and can be owned. It does not acknowledge that resources bear a relationship to each other in the natural environment as part of environmental systems. Market mechanisms fail to allocate environmental goods and services efficiently precisely because environmental systems are not divisible,

16

frequently do not reach equilibrium positions and incur changes which are not reversible... Economics is not adapted to consider total changes. Resting as it does on the concept of the margin, it is epistemologically predisposed to a reductionist view of resources and utility... Ecosystems are themselves a source of value (p.40-41).

It is amongst 'dark-greens' (Dobson, 1989, p.89) and the 'Deep Ecology' school (Dobson, 1989, p.47, Redclift, 1987, p.43) that the value of eco-systems is more seriously addressed. According to Redclift, the underlying presumption of the latter paradigm is biocentric (earth-centred) rather than anthropocentric (human-centred), emphasizing human beings' underlying unity with other living beings and processes. Although there are many different positions within this school and, indeed, with other dark-greens, what normally is shared is a deep scepticism of current industrial processes and values and an opposition to economic growth per se. Daly (1980, p.325), for example, projects a 'steady-state economy' where the 'throughput of matter-energy ...is reduced to the lowest feasible level'. Meanwhile, Porritt (1989) argues that 'a low energy strategy means a low-consumption economy; we can do more with less, but we'd be better off doing less with less' (p.174) and that 'progress in the future may consist in finding ways of *reducing* GNP (p.121, emphasis added).'

There are many difficulties with a zero-growth position, not least being that it holds no political appeal amongst those whose current GNP per head is low. But there are also logical problems. Jacobs (1991, p.54) defends the concept of economic growth from those who take the zero growth option:

> Unfortunately, the 'growth debate' has generated much more heat than light. Indeed, in many ways it has distracted attention from the real issues of the environmental crisis. By linking environmental concern to 'zero growth' it has allowed those dismissive of this position consequently (but irrationally) to dismiss the environmental problems it seeks to address.... Because current patterns of economic growth are environmentally damaging, it does not follow that the solution to environmental problems is no growth.

For Jacobs (1991), the crucial concept is 'the environmental impact coefficient' (EIC) of GNP which he defines as 'the degree of impact (or amount of environmental consumption) caused by an increase of one unit of national income'(p.54). If the EIC falls faster than GNP rises, then the overall impact of growth on the environment falls. There are, according to Lecomber (1975, p.42), three systematic changes in production processes which can theoretically reduce the EIC as GNP rises: (1) a change in the composition of final output towards less environmentally damaging products (e.g. from goods to services), (2) the substitution of less environmentally damaging factor inputs for more damaging ones (e.g. from fossil fuels to renewables) and (3) an increase in the efficiency of resource use through technological progress (e.g. by increasing energy conservation efficiency). Should they take place, these processes, known by the World Bank as 'substitution, technical progress and structural change' (World Bank, 1992a, p.9), allow the possibility of an increase in GNP to be accompanied by a reduction in negative environmental impacts. The discussion is taken up further by Ekins (1993, p.281) and Ekins and Jacobs (1995, pp.25-29). The 'all-important

17

equation' (Ehrlich and Ehrlich, 1990, p.228) relating environmental impact and human activity is written:

$$I = PCT$$

where I is environmental impact, P is population, C is consumption per head and T is impact per unit of consumption (Jacobs' EIC). This identity makes it clear that if growth is to occur without environmental constraint then EIC must fall and that should P or C rise (which is likely to be the case, particularly in developing countries), then for any given reduction in I, the EIC must fall all the further. However, since it is important that the reduction in EIC is not brought about by increasing intermediate costs and hence the capital-output ratio, the net costs of doing so must be neutral or negative. To these conditions for environmentally costless growth, Ekins and Jacobs (1995, pp.30-1) add three others, that reductions in EIC must be continuous and exponential, they should occur across all relevant environmental impacts and the growth-retarding effects of any *prior* negative environmental impacts should be cleared up simultaneously with current impacts.

Ekins and Jacobs (1995, p.29) discuss two further circumstances in which reducing the EIC will itself lead to productivity gains, and thus promote increases in GNP, first when governments discontinue policies which are both environmentally unfriendly and economically inefficient, and secondly when businesses introduce changes in methods, products or processes which are simultaneously environmentally beneficial and cost saving. Both cases lead to a 'win-win' situation (World Bank, 1992a, p.1).

Thus Ekins and Jacobs argue that so long as the above conditions concerning the fall in EIC hold, it is *theoretically* possible to increase GNP indefinitely without hitting an environmental constraint. However, 'the theoretical possibility does not guarantee its practical achievability' (Ekins and Jacobs,1995,p.31). As Lecomber (1975,p.42) has put it, this:

> establishes the logical conceivability, not the certainty, probability or even the possibility in practice of growth continuing indefinitely. Everything hinges on the rate of technical progress, and possibilities of substitution. This is perhaps the main issue which separates the resource optimists and resource pessimists. The optimist believes in the power of human inventiveness to solve whatever problems are thrown in its way, as apparently it has done in the past. The pessimist questions the success of these past technological solutions and fears that future problems may be more intractable.

In summary, with regard to the possibilities of sustainable development resulting from economic growth, the argument of this chapter that different ideological paradigms can be categorized within an optimistic/pessimistic framework as suggested above. Into the category of policy optimists go neo-classical economists, Pearce et al., the World Bank (see World Bank 1992a) and members of the Brundland Commission, amongst the pessimists go orthodox Marxists, Deep Ecologists and such scholars as Mishan, Daly and Meadows et al. Agnostics include political economists such as Redclift, Ekins and Jacobs. In the latters' words, (1995, pp.41-2),

It is clear that as far as environmentally sustainable G(N)P growth is concerned, there is everything to play for but the going will be exceptionally tough... It is certain that the currently dominant business-as-usual approach, going for G(N)P growth with a few environmental add-ons, will not address the gathering environmental crisis.

It is clear that welfare can in some circumstances be increased without any increase in measured GNP, a more egalitarian redistribution of income might well achieve this as might some increase in the 'feel good factor' not associated with material things. But to the extent that policy makers tend 'to go for GNP growth' to achieve an increase in welfare, this thesis belongs squarely within the agnostics camp as described above.

2.10 Sustainable Agriculture

Economic development brings a range of problems which seriously threaten the sustainability of life and livelihood and this is as true of modern agricultural developments in poor countries as of recent industrial change. Clearly, all agriculture takes some toll of the immediate ecological balance and 'environmental management' is not new: traditional agricultural practices in China itself, for example, have for centuries highlighted the importance of ecological balance (King, 1926, Catton, 1992). However, developments in the post-war years, particularly the globalisation of agricultural production have led to significant changes both in agricultural forms and techniques with the primary intention of maximising agricultural output to earn increasingly large amounts of cash. These changes have been successful in so doing and have contributed to a significant increase in global food production and a welcome reduction in hunger and malnutrition. But at the same time they pose threats to ecological sustainability over time .

Focusing on the latter, Bartelmus (1986, p.44) argues successful 'evolutionary adjustment' of ecosystems (Odum, 1971, p.35) to new patterns of resource use typically demands high diversity, large biomass and high stability, and that mature ecosystems displaying those characteristics, such as tropical rain forests, achieve stability when part of the energy flows within the system are directed towards the maintenance of the system itself, rather than towards production. As Bartelmus (1986, p.44) writes,

> In young systems the rate of gross production of biomass and organic matter tend to exceed the rate of community respiration, that is the maintenance costs of the ecosystem. Mature systems on the other hand exhibit equal or near-equal rates of production and respiration.

Attempts to maximise agricultural production threaten ecological succession, therefore, because it ceases to be in the interests of agriculturalists to maintain mature systems. Instead, their best interests are served by maintaining artificially young productive ecosystems where organic matter and biomass are not allowed to accumulate, thus achieving higher levels of short-run productivity at the expense of natural means of protection and adaptability. Furthermore, as Redclift argues (1987, p.18):

19

In addition, the maintenance of an ecosystem in an artificially young state of productivity, under crop production for example, requires enormous energy subsidies in a form not available in nature. Fertilisers, fuels for machinery, irrigation technology, genetic selection of species and pest control, are all facets of this attempt to renew immature ecological systems in a state of high productivity...In essence, agricultural development implies a necessary threat to ecological succession, in which costly energy subsidies replace natural processes.

Conway (1985, p.35) defines four desirable properties of agro-systems - productivity (net output per unit of resource), stability (the degree to which productivity is constant in the face of small disturbances, e.g. climatic fluctuations), sustainability (the degree to which productivity can be maintained in the face of a large disturbance, e.g. serious soil erosion) and equitability (the degree to which the benefits of output are evenly enjoyed amongst villagers) and stresses the ways in which methods of cultivation highly productive in the short run pose threats to stability and sustainability, highlighting the double-edged results of the Green Revolution. Conway (1985, p.35) writes,

Traditional agricultural systems ...have low productivity and stability, but high equitability and sustainability. However, the introduction of new technology, while greatly increasing the productivity, is also likely to lead to lower values of the other properties. This was particularly true, for example, of the introduction of the new high yielding rice varieties, such as IR8 and its relatives, in the 1960s; yields fluctuated wildly, but have tended to decline, in part due to growing pest and disease attack. More recent varieties combine high productivity with high stability, but still have poor sustainability.

Soil erosion is potentially the most immediate livelihood threatening result of intensive agriculture, making soil less able to retain water, reducing its depth and depleting it of nutrients. Moreover, eroded topsoil can be carried into lakes and reservoirs, silting up waterways, ruining fisheries and increasing the incidence and severity of floods. Soil degradation further reduces the resource base by encouraging the use and overuse of marginal lands and the conversion of forests for food cultivation in turn frequently leads to destructive deforestation, bringing with it the loss of livelihoods and communities for forest dwellers, a further loss of soil protection and enrichment and in mountainous regions the disruption of upland watersheds and the ecosystems that depend on them, further threatening increases in the incidence and severity of floods and drought. Deforestation also has more global, life-threatening impacts, contributing to loss of biodiversity and accelerating the 'greenhouse effect'.

Chemical fertilisers are used to boost soil nutrition in the short run, but there is considerable danger to water resources as a result of nitrogen and phosphate run-off while in the long-run, diminishing returns set in, requiring more and more chemicals to maintain existing levels of output. The use of pesticides can be life-threatening: as a result of long-run exposure to chemical and pesticide residues in food and water, there are all too many deaths in poor countries, as Rachel Carson reminded us 35 years ago (Carson, 1962).

Irrigation has expanded dramatically, but its legacy has, in recent years, frequently been salinization and waterlogging which has significantly reduced the expected increases in productivity of irrigation investments. Meanwhile, inappropriate and inefficient irrigation methods have contributed to one of the most livelihood threatening of all environmental problems, particularly in China, drops in the water table and water shortages.

The most manifestly life and livelihood threatening effects of inappropriate economic practices in rural areas has been the process of desertification. According to the Brundlandt Commission (WCED, 1987,p.127), in 1987, the world's drylands supported 850m. people, some 230m. of whom lived on land severely affected by desertification. Meanwhile land degraded to desert-like conditions, or providing no possibilities for eking out any kind of livelihood, were continuing to grow at a rate of 27m. hectares a year.

There are other disturbing features of recent agricultural developments, irrespective of their sustainability as discussed above. As Redclift points out, the productivity of modern agricultural methods when compared to traditional ones is judged most frequently in terms of the productivity of labour or land, rarely in terms of energy efficiency (1987, p.22). Yet work done by Gligo for the UNEP in 1985 (Redclift, 1987, p.27) concluded that from the standpoint of both energy efficiency and the productivity of the agricultural system (in energy terms) per unit of land, the most successful systems were those combining crop rotation and fallow with low energy inputs, such as experienced in many traditional farming systems. Given the financial costs of oil-based energy and the implications of increased dependence of many developing countries on imported oil, coupled with all the problems that energy generation in general terms poses for the environment, the benefits of modern methods in agriculture to the development process are put in doubt. As Redclift (1987, p.29) concludes,

> For the less developed countries, one of the most potent arguments against using more oil-based energy is that apart from its financial cost and ecological effects, such energy feeds technological practices that make agriculture less rather than more efficient.

The discussion so far has concentrated on the concept of the *ecological sustainability* of agricultural systems. The consideration of such sustainability is, according to IUCN (1991, p.4) 'essential, because, regardless of short-term gains, *productivity without sustainability is mining.*'

However, the sustainability of agricultural systems does not rest purely upon ecological sustainability. As Susannah B. Heckt (1987, p.5) argues,

> Agro-systems have various degrees of of resiliency and stability, but these are not strictly determined by biotic or environmental factors. Social factors such as a collapse in market prices or changes in land tenure can disrupt agricultural systems as decisively as drought, pest outbreak or soil nutrient decline.... The results of the interplay between endogenous biological and environmental features of the agricultural field and exogenous and social and economic factors, generate the particular agroecosystem structure.

It is clear that for agriculture to be *sustainable,* therefore, it is necessary to look beyond ecology and, indeed beyond technology, towards political, economic and social relations. For

21

Blaikie and Brookfield (1987, p.1), land degradation is 'a social problem' while in Redclift and Sage's words (1994, p.10), 'agricultural policies have environmental impacts that are mediated by human agents'. As Cai and Smit emphasize (1994, p.300), not only does agriculture operate within a biophysical environment but within socio-political and economic and technological environments. For Cai and Smit (1994, p.300-1),

> The socio-political environment is concerned with the role that human populations, cultural and other forces of social and collective action play in influencing the behaviour of people as individuals, and as members of families, groups and communities.... For agriculture to be sustainable it must be compatible with the socio-political environment within which it operates.

Meanwhile,

> The economic and technological environment greatly constrains the feasibility or viability of agricultural activities.....Without the likelihood of sufficient returns to at least cover costs of production, operators lack incentives and ultimately the ability for engaging in agriculture. If agriculture is not undertaken, it is hardly sustainable (Cai and Smit, 1994, p.301).

The work of such scholars as Blaikie and Brookfield (1987), Redclift (1987), Tisdell (1988), Lele (1991) and Cai and Smit (1994) all point to the difficulties involved in understanding sustainable development in general and sustainable agriculture in particular in abstract from a political-economic analysis. Clearly, it is very important to develop technically feasible agricultural systems which are friendly towards the natural environment but whether those technically feasible systems are adopted are almost exclusively to do with those political-economic, social, factors which determine whether farmers are predisposed towards their adoption and use. As Cai and Smit emphasise (1994, p.301),

> There are few places where agriculture is not greatly influenced either directly or indirectly by the policies and actions of governments... In some countries, the political environment is reflected in production quotas, distinctive forms of land tenure, controls on inputs and so on. Thus, for agriculture to be sustainable it must be compatible with the socio-political environment in which it operates. Such compatibility, and viability of agriculture, may be achieved by adjusting the agricultural activities *or by a change in the socio-political conditions* (emphasis added).

In a country like China, where the *political* so strongly and overtly determines what is and what is not possible at every level of analysis, it is essential that a thorough-going political-economic analysis of potentially viable systems be made, hence the need for the political economy of CEA undertaken in this book.

Given the variety of academic disciplines from which the literature on sustainable agriculture springs (ecology, agronomy, economics, political economy, geography) it is unsurprising that different scholars lay emphases on different aspects of sustainability. As suggested above, Conway (1985) stresses *ecological sustainability,* judging sustainability primarily in terms of the ability of an agricultural system to withstand and recover from stress, while Tisdell (1988, p.374) though accepting the ecological constraints puts more emphasis on 'sustainable patterns of economic exchange as well as sustainability of policies and social structures and sustainability of community'. Dasmann (1984, see Cai & Smit, 1994, p.303) judges sustainable agriculture additionally in terms of the fulfilment of basic needs and self-reliance, while Altieri (1989, see Cai & Smit, 1994, p.303) focuses on sustainability in terms of land restorativeness, environmental soundness, economic viability and social acceptability. All scholars are agreed, however, that sustainability in the countryside has social, political and economic as well as ecological dimensions.

2.11 Conclusion

Thus sustainable agriculture has a variety of different dimensions and can operate on a number of different geographical scales and through different time periods. Of course, many questions may still remain over precisely what is to be sustained and still more over the methods of achieving it. But, mindful of those problems and of the discussions above, this thesis will accept a working definition of rural sustainability as *development in the countryside which threatens neither life nor livelihood in the medium or long run* and will examine it with reference to the Chinese countryside in the context not merely of ecological constraints but also of economic, political and social relations.

To the extent that there are no easily quantifiable macro-economic measures of rural sustainability which are directly applicable to a micro-economic analysis of specific environmental impacts in Chinese villages, this work will ultimately be prosecuted conceptually and qualitatively. With regard to the applicability of the above definition of rural sustainability to the Chinese countryside, the sustainability of CEA will be judged in terms of its ability to provide the basis for increases in welfare, viz. in increases in material benefits and the general *quality of life* for present villagers without compromising the opportunities for future generations to enjoy similar benefits. Chinese literature frequently judges the outcomes of newly adopted initiatives such as CEA in terms of: (i) economic benefit, (ii) social benefit and (iii) environmental benefit. Despite the somewhat arbitrary and sometimes awkward nature of this classification, I have nonetheless adopted it in my conclusions on the impacts of CEA in the villages studied, reflecting the fact that material benefits alone, without positive environmental impacts, are likely to be short-lived and ultimately unsustainable as livelihoods are destroyed and local economies regress.

3 Developments in the Chinese Rural Political Economy 1949-1998

3.1 Introduction

On October 1 1949, Mao Zedong pronounced the birth of the Peoples' Republic of China from the Gate of Heavenly Peace in Tiananmen Square, Beijing. In the forty years that followed, China experienced unusually turbulent social change. Though ruled for the entire period by one party, the Chinese Communist Party, the people were subjected to a bewildering series of social experiments, as one wing of the Party or another gained ascendancy and differentially attempted to create, sustain, reform or radicalize society in the hopes of transforming socialist theory into political practice. But there was *some* consistency: throughout the period and despite differences in emphasis (see Yao, 1994, p.4), the Chinese countryside was squeezed of surpluses in the interests of the heavy industrial and urban economies. As White (1993, p.96) writes,

> State policies towards agriculture.. were not designed to encourage rural development merely for its own sake, but to link agriculture and other sectors of the rural economy into an overall economic strategy which benefited industrialization.

3.2 The Rural Political Economy of China 1949 to 1978

3.2.1 The pre-liberation rural political economy

In 1949, China had a population of 541.7 millions, 89 per cent of which lived in the countryside (China Statistical Yearbook, 1992, p.63), and the great majority of them were farmers living, given the physical geography of China, on one-sixth of the total land area (Riskin, 1987, p.22-23). Pre-liberation farming was dominated by crop-growing, with 90 per cent of the total farm area devoted to crops and only 1.1 per cent given over to pastureland for animals (Buck, 1964, p.268) and Chinese farms were very small, with 70 per cent of all farms less than one hectare, the median size being 0.97 hectares, or approximately 16 *mu*. However the small-scale of farming was magnified further by partible inheritance and free commerce in land which led to farm fragmentation, the typical farm being made up of an average of 6 plots, at a mean distance of 0.4 miles from the farmhouse. This led not only to land wastage because of the need to demarcate boundaries and maintain footpaths, but also to a waste of labour time and effort, particularly in the busy season when farmers had to move themselves, their equipment and animals from field to field. Furthermore, fragmentation rendered irrigation and drainage difficult and led to frequent disputes over ownership and access (Riskin, 1987, p.23).

Different studies paint different pictures with regard to the social structure and distribution of income (see Riskin, 1987, pp.28-9, Chossudovsky, 1986, pp.26-7), but all point to a majority of households owning little or no land. Liu's study (1953, p.19, cited in Riskin, 1987, p.28), for example, suggests that 10 per cent of all households were entirely landless, a further 17.3 per cent owned less than 5 *mu* and a further 40.2 per cent owned between 5 and 20 *mu,* these households between them making up 68 per cent of all households, yet owning only 24 per cent of the total cultivable area. On the other hand, the richest 11 per cent of households owned 44 per cent of the cultivable area. What social mobility existed, moreover, was, largely because of partible inheritance, in a downward direction. The incidence of landlordism varied from place to place; in the areas where it was most prevalent, landlords were largely absentees, living in towns and cities and siphoning off income for commercial or speculative purposes although in others, little land was rented out and the landlords were hardly differentiated from other social classes. But whatever the particular circumstances, Riskin argues that socio-economic conditions in the pre-liberation Chinese countryside were a major deterrent to investment, innovation and growth:

> High rents and increasingly insecure tenancy discouraged investment by tenants, while fragmentation of holdings discouraged technical improvements. Where commercialization might have stimulated investment, innovation and specialization, inadequate credit and high interest rates made borrowing for such purposes a major gamble; few peasants were in a position to undertake it, or were willing to trust their fate to an unpredictable market (1987, p.32).

3.2.2 Land reform

In 1936, Mao proclaimed 'whoever wins the peasants will win China, whoever solves the land question will win the peasants (quoted in Snow, 1981, p.49).' The 'Yan'An Way' was the rural strategy adopted by Mao after the Long March of 1934-5, which, while stressing firm Party leadership tempered by the 'mass line', involved principles of co-operation based on self-management, private organisation, voluntarism and mutual benefit (Howard, 1988, pp.23-26) and gradualist policies including support for rent reduction rather than land reform. But with the intensification of the civil war between the Communists and the Guomindang Mao needed to win the peasantry more firmly to his side. Thus the Land Reform Law of September 1947 was promulgated to buy the peasants' support, demanding the confiscation of all land owned by the (Guomindang) state and the redistribution of any land above the average, including that of the rich peasants as well as landlords, to poor and landless peasants (Brugger and Regler, 1994, p.13). In the following winter and spring, poor peasants in many areas carried out 'struggle' and 'settle accounts' meetings, leading frequently to violence against and execution of landlords, and the resulting 'local commandism' and chaos was only brought to an end with the passage of a more moderate land reform law in 1949 and the imposition of stricter discipline from above (Riskin, 1987, p.52; Brugger and Reglar, 1994, p.95). From 1949 on, a more cautious approach to land reform, dubbed 'the rich peasants line' (see Chossudovsky, 1986, p.30), was taken in the newly liberated areas, mostly in southern and central China, but pressure from

China's involvement in the Korean War hastened the process so that, towards its end, land reform was being carried through at great speed. By the end of the movement, in 1952-3,

> landlords, estimated nationally at some four to five per cent of the rural population, had lost their former land with some 400,000-800,000 killed (one in six landlord families) and 44% of cultivated land had been distributed (Brugger and Reglar, 1994, p.96)

3.2.3 The mid-1950s: consolidation and co-operatization

Land reform created a new political economy in the countryside based on small-scale cultivation of land individually owned and farmed. Using the categories introduced by the 1950 Agrarian Reform Act (Riskin, 1987, p.31), the average size of the holdings of poor peasants' was 0.8 hectares (13 *mu*), of middle peasants' 1.3 hectares (21 *mu*) and of rich peasants 1.7 hectares (27 *mu*), too small in most cases, therefore, to provide for a sufficient calorie intake still less a surplus for investment (Brugger and Reglar, 1994, p.97). Thus it is unsurprizing that this small-scale, effectively privatized agriculture, did not survive for long as the authorities required a more productive agriculture from which to squeeze the surpluses on which the planned growth of the industrial sector during the years of the 1st Five Year Plan (1953-57) was based. The mid 1950s, therefore, saw a period of accelerating cooperatization in three main forms: the consolidation of 'mutual-aid teams', composed of four or five neighbouring households, pooling tools, animals and effort during the busy season, of 'elementary cooperatives', composed of twenty to thirty households who combined assets and earned income in two ways, either in dividend payments for land, animals or tools and renumeration for work performed, and of 'advanced cooperatives', initially of about thirty households which pooled all means of production and earned income based on labour contribution via a work-points system (Lin, 1990, p.153).

The government encouraged the various producer cooperatives to become the purchasing and sale contractors with the newly created state-organized supply and marketing cooperatives. By 1954, 114,000 cooperatives had been formed (Brugger and Reglar, 1994, p.98). Considerable difficulties were involved in forming and holding them together, particularly as a result of managerial diseconomies and the lack of incentives for farmers to work as diligently for the collective as they had done before on their own private plots (Nolan, 1990, p.6) but they encountered no *active* resistance from the peasantry (Lin, 1990, p.153). Thus, in 1955 Mao announced a new strategy for agriculture which involved still more ambitious production targets and an acceleration in the co-operatization movement. By the end of that year, as the movement reached its 'socialist high tide', 75 million peasant households, approximately two-thirds of all peasants, had joined co-operatives.

3.2.4 The Great Leap Forward (da yue Jjn) 1958-1959 and its aftermath

The gradualist co-operatization policy accepted by the Party leadership in 1956 and 1957 was interrupted on November 13 1957 when the *Peoples' Daily* published an editorial entitled 'Start a New Upsurge in Agricultural Production', calling on people to 'make a *great leap forward* on the production front' (Party History Research Centre of the CCP, 1991, p.268, emphasis added). This was the first time that the slogan 'Great Leap Forward' emerged, one to be repeated many times in the following year as Mao determined that China should catch up with Britain, then the second largest producer of steel in the world, within ten years.

The Great Leap Forward had enormous implications throughout Chinese society, but the greatest changes took place in the countryside. Confident of the benefits of economies of scale in agriculture, Mao encouraged the amalgamation of the collective farms, which by 1957 numbered 700,000, into 24,000 rural peoples' communes comprising an average of over 22,000 peasants, with centralized accounting and work organization (Nolan, 1988, p.6). And huge changes took place in all aspects of rural socio-economic life:

> the private plot was virtually eliminated; cooking eating, laundry services and childcare were frequently socialized; income differentials between villages within the commune were removed; a large part (in some cases all) of the collective income available for distribution to households was distributed according to the 'communist' principle of 'to each according to their need' (Nolan, 1988, p.7).

Though the average size of communes was substantially reduced (the 26,000 communes were sub-divided into 74,000 communes [Aziz, 1978, p.15]) the commune *system* which emerged after the Great Leap Forward remained essentially unchanged until the reforms of the late 1970s and early 1980s, consisting of three 'nested layers' (White, 1993, p.95), the commune *(gongshe)* the brigade *(dadui)* and the production team *(xiaodui)*. The commune comprised between ten and twenty brigades, while each brigade comprised between five and ten teams. The commune, initially with over 20,000 people, and often composed of a small market town and surrounding villages, was responsible for all major Party, welfare, infrastructure and planning services, including major rural investment projects such as irrigation, land reclamation and agricultural extension. The brigade was responsible for small-scale rural infrastructure, primary schools, clinics, and local industry, and was the site of the lowest branch of the Party (White, 1993, p.96), while the team, comprising twenty or so neighbouring households operated as the basic unit of agricultural production. It formally owned the land cultivated by its members and was the basic 'unit of account'; the income of each working member was his/her share of the team's total income net of taxes to the higher authorities for social welfare, infrastructure and so on, of debt payments and of payments to a sinking or accumulation fund at the brigade level for longer-term investment projects. Remuneration was measured in terms of 'work-points' awarded to each worker on the basis of the time taken and the importance or difficulty of the task. Individual incomes varied, therefore, not only on the size of the harvest, on the amounts creamed off for collective investment and welfare and on the level of procurement prices and subsidies set by the state (White, 1993, p.96), but also on the degree of collective effort - any one worker's income depended to some extent on the efforts of others.

While the institutional changes to the political economy of rural China were perhaps the most profound and long-lasting of the elements of the 'Great Leap Forward', the most spectacular

27

was the attempt to boost steel production in the countryside by melting down scrap-iron in small, back-yard blast furnaces, fired with charcoal from trees indiscriminately felled, demanding enormous human exertion as the work was all done by hand (Smil, 1987, p.217). But the results of the 'Great Leap Forward' were not as Mao had hoped. It is now generally agreed, even within orthodox CCP circles, that the movement was a spectacular failure.

While there may be some measure of disagreement as to the precise meaning of 'sustainable development', there can be little dispute that the 'Great Leap Forward' represented a monument to development of a wholly *unsustainable* nature. The ecological environment was not spared with millions of trees felled to provide fuel for the blast furnaces, 'exacerbating ecological problems in an already denuded countryside (Howard, 1988, p.33).' Qu Geping comments that 'the result was detrimental to the environment. Today, we are still suffering from the bad effects of that period' (Catton, 1993, p.6).

Combined with ecological difficulties, the political economy of the countryside was unable to withstand a couple of difficult winters, with poor harvests in 1959 and 1960 camouflaged by a sea of official exaggeration and enforced transfer of food to the urban areas. Endemic famine in the countryside struck in 1960. Whatever the true size of the death toll resulting from the Great Leap (official estimates suggest 10 millions, other scholars suggest three times as much) there is no question that 'the collapse of agricultural production was accompanied by a demographic disaster' (Nolan, 1988, p.49). Apart from the number of deaths, the number of foregone or postponed births is estimated at 33 million (Smil, 1987, p.217).

The aftermath of the Great Leap Forward led to official retrenchment and a 'New Economic Policy' in 1961 which, upholding the slogan of the 'three freedoms and one contract'*(sanzi yibao)* restored private ownership and self-management rights, (households were allowed to earn up to 20 per cent of their income from private sideline activity) abandoned compulsory transfer, revived rural markets, reduced the size of the communes and fixed output quotas on individual households (Howard, 1988, p.41; Brugger and Reglar, 1994, p.112). There was some immediate revival of the rural economy, if at some social cost as schools and clinics previously supported by collective funds collapsed. But in 1962, even as the 'gradualist' rural economic revival began, Mao launched a new mass movement, the Socialist Education Movement to weed out any capitalist tendencies amongst villagers, emphasizing the importance of 'taking class struggle as the key link' (Party History Research Centre of the CCP,1991, p.306) and insisting that 'class struggle must be stressed every year, every month and every day' (p.301).

In February, 1964, the 'Movement to Learn from Dazhai' was launched. Dazhai was a production brigade in the mountainous regions of Xiyang County, Shanxi Province that had significantly boosted farmland capital construction and increased agricultural output through co-operative activity, the three main characteristics of the movement being the stress on 'class struggle', the importance of the brigade rather than the team as the accounting unit, and the significance of a particular form of work-point system 'known by the masses as "political work-points" rather than "labour work-points"' (Wang et al., 1985, p.39). In so doing it put stress on larger rather than smaller units and called for a greater degree of egalitarianism. (Howard, 1988, p.42).

The 'Socialist Education Movement' and the 'Movement to Learn from Dazhai', however, were merely the forerunners to Mao's more ambitious movement to nip in the bud any capitalist retrenchment in the countryside, a movement as much waged against perceived bureaucratization in the Party as against anything else, the Great Proletarian Cultural Revolution (wenhua da geming).

3.2.5 The Cultural Revolution 1966-1976

The immediate results of the Cultural Revolution in the countryside involved the elimination or collectivization of all private plots, the strict discouragement of side-line activity in order to cut off any 'capitalist tails' (ziben zhuyi weiba), the restriction of rural free markets and the reinforcement of the brigade as the key production and accounting unit. There was considerable chaos, particularly in the first two years, as old cadres were frequently subjected to violent criticism (piping), denounced and displaced by revolutionary committees, insisting upon the primacy of class struggle over production. Despite this, however, the Cultural Revolution did not disrupt the rural economy to the same extent as the Great Leap had done. Indeed, there were some gains associated with the recollectivization involved. Labour, for example could be more easily mobilized to produce 'public goods'. During the Cultural Revolution, the peasants built over 80,000 reservoirs and laid irrigation systems on 76.7 million acres of land, largely by hand. Meanwhile, the communes provided a network for rural health services, co-operative medical insurance schemes, a very large number of primary schools, roads, power stations and an electricity supply system to many villages (Howard, 1988, p.43). These things were achieved at a very high price, however.

The environment was not spared: the Cultural Revolution is viewed by most scholars as a period of ecological disaster, particularly as it was more important to be red than expert and the training and quality of forestry workers consequently suffered. (Edmonds, 1994, p.48) Individual households lost control over the decision-making process. Peasants were obligated to sell quotas to the state at low prices, thus effectively subsidizing town dwellers, while they often had to pay high prices for their manufactured inputs, such as chemical fertilisers, thus finding themselves badly mauled by the notorious 'price scissors'. Meanwhile sideline activities were condemned as 'capitalist tails': even the making of cakes or straw hats or the keeping of pigs was frequently outlawed, thus eliminating outlets for spare-time labour, extra sources of cash and availability of local goods and services. Farmers lost individual ownership rights over their means of production including their land, animals, carts and tools and were effectively 'proletarianized' while the state dictated to them as a 'collective capitalist' (Howard, 1988, p.44).

The worst excesses of the Cultural Revolution in the countryside and elsewhere took place in its first two or three years (1966-68) and by the early 1970s there was a time of retrenchment of the three-tier commune-brigade-team structure in a period of general stability. However, the Cultural Revolution came formally to an end only after the deaths of Premier Zhou Enlai and Mao Zedong in 1976, with the downfall of the Gang of Four on October 6 of that year. In the aftermath Deng Xiaoping slowly but surely wrested power from other sections of the Party hierarchy and in 1978 initiated a series of historic reforms beginning in the countryside.

3.2.6 An overview of development in the rural political economy 1949-1978

By the end of the 1970s, the Chinese government could boast of progress in a number of different aspects of the rural political economy. Gross output went up every year from 1952 to 1978 in real terms (China Statistical Yearbook 1992, p.299) and between the same years, the number of large and medium sized tractors increased from 1,307 to 557,358, of combine harvesters from 284 to 18,987, of trucks for agricultural use from 280 to 97,105, (China Statistical Yearbook, 1992, p.304), the amount of tractor-ploughed land increased from 136,000 hectares to 40,670,000 hectares, the area of irrigated land from 19,959,000 hectares to 44,965,000 hectares and the consumption of chemical fertilisers from 78,000 tons to 8,840,000 tons (China Statistical Yearbook, 1992, p. 312). Meanwhile, China enjoyed a high degree of equality, the World Bank's estimate of the Gini Coefficient for 1979 being 0.26 (Brugger and Reglar, 1994, p.126) and its ability to meet the basic needs, particularly health needs of the vast majority of the population and the high level of life expectancy - by 1979 it was 64, the highest for a developing country in World Bank Statistics (World Bank, 1991, pp.168-178), and up from 40 in the early 1950s (Smil, 1993, p.17), were impressive. The communes had mobilized labour which had, frequently by hand, constructed public goods from dams to reservoirs to power stations. There are scholars (e.g. Muldavin, 1992, Hinton, 1990, Chossudovsky, 1986) who believe that the achievements of the Maoist period have been downgraded by an 'opportunist' (Chossudovsky, 1986, p.4) alliance of western 'scholars' and the CCP leadership. It is difficult not to agree with Howard (1988, p.43), however, that those achievements were bought at a high price.

Traditional peasants had been transformed into 'socialist' landless workers (Zhou, 1996a, p.2). But there were no spectacular productivity gains. Indeed, after 1956, the output of many staple crops did not keep up with population growth. The per capita grain output *fell* by 3.6 per cent between 1956 and 1977, the per capita cotton output *fell* by 0.2 kilograms and the per capita output of oil-bearing crops *fell* by 1.2 kilograms. Between 1965 and 1976, farmers' per capita income from workpoints had increased by only 10.5 yuan, representing an increase of less than one yuan a year, just enough to buy a packet of cigarettes, and private earnings from sidelines had been seriously reduced. In 1978, per capita earnings from collective sources averaged only 74 yuan and a quarter of all peasant households had a per capita income of less than 50 yuan (Howard, 1988, p.44). Most farmers still lived in houses made of mud and straw, cooked and slept on the traditional *kang,* had no indoor sanitation or electricity, had next to no consumer durables and ate little except rice. From the government's point of view the situation was entirely unsatisfactory, from the farmers' point of view it was a great deal worse. As Howard (1988, p.44) puts it:

> Peasants succinctly expressed their predicament with a common complaint that they were being 'roped together to live a poor life.' It was time to cut the rope.

3.3 The Political Economy of the Chinese Countryside 1978-1996

3.3.1 Deng Xiaoping and the agricultural reforms of December 1978

When Deng Xiaoping finally came back to power in 1977, he had an agenda with goals to:

> get the party to repudiate the ideology, policies and purges of the Cultural Revolution, *to substitute development for class warfare* as the highest order of business and to allow bold economic experiment (Evans, 1993, p.228) [emphasis added].

In common with early Mao, Deng and his influential colleague in the politburo, Chen Yun, understood the importance of balanced growth ('walking on two legs'), specifically the importance of a strong and vigorous agricultural sector as support for the industrial sector. Both understood that the fundamental problem of the Chinese economy was low agricultural incentives and excessive state investment skewed to heavy industry, which encouraged the rapid expansion of the industrial workforce and pressure to feed them cheaply. This problem was compounded in 1980 when Deng mused publicly that it would be possible to quadruple *(fan liang fan)* per capita output by the year 2000 (Far Eastern Economic Review, April 28, 1983,p.44). According to Kelliher (1992, p.26), agriculture could not carry the burdens because,

> incentives for farmers to produce the surplus demanded by the state were too small. Farm-good procurement prices were set too low. Furthermore, central planning was so heavy handed and inflexible that it prevented farmers from pursuing the subsidiary production that might have elicited their enthusiasm. And finally, light industry had been so slighted, farmers had little reason to want to make more money even if they could: there were hardly any consumer goods to spend it on.

Turning his back on the Stalinist approaches to agriculture which Mao had adopted, and taking the Bukharinist path (Nolan, 1988, ch. 1) Chen Yun had a bold strategy for agriculture for which Deng won total support at the 3rd Plenum of the 11th Central Committee of the CCP in December 1978. It included shifting state investment from heavy into light industry and into agriculture, higher farm procurement prices, a decentralization of the basic accounting unit from the brigade to teams or groups of households, a switch from output planning to price planning, using relative price differentials to allow farmers to plant more lucrative crops suited to local conditions, the reintroduction of private plots within communes abandoned during the Great Leap Forward, (see Croll, 1983, p.158) and the abolition of many of the restrictions on sideline activities. But he was preempted by the farmers themselves (Kelliher, 1992, p.17), as spontaneous attempts to return to family farming broke out in the north, centre and east of Anhui province. In Xiaogang Brigade, Fengyang County, for example, suffering from unusually harsh poverty as a result of drought, the farmers made a decision one night in mid-1978:

> We secretly decided to distribute the land to each household, which in turn promised to deliver the grain tax for which they were responsible and not ask for help from the state. We were prepared for the worst to

31

be jailed or sentenced to death for doing this. The other commune
members agreed to raise our children until they were eighteen years old
(China Pictorial Publications Co.(Ed.), 1989, p.12).

Similar decisions were made in Shannan commune, Feixi County and Huangchi commune in Wuhu County (Kelliher, 1992, p.61). The decisions had to be clandestine because the newly established systems effectively involved the principle of fixing farm output quotas for each household, the very contract system of the *sanzi yibao* period in the New Economic Policy of the early 1960s, outlawed in the Socialist Education Movement and Cultural Revolution, and thus regarded as 'anti-socialist'. But, fortunately for the farmers and local cadres in Anhui province, the Party First Secretary *shuji* Wan Li gave public support to the local initiatives in his province. Communes in Sichuan province began similar moves and found tacit if not active support for them from the then provincial Party First Secretary *shuji* Zhao Ziyang, by 1980 to become national premier, who formulated a 'four don'ts' *(si bu)* policy to deal with them: don't publicize them, don't oppose them, don't support them, but don't stop them (Kelliher, 1992, p.79).

While the central government leadership was keen to decentralize the production decisions from the brigade to teams of groups of households *(bao chan dao zu)* , the return to household contracting was originally opposed and in its opposition they had the support of most local cadres whose personal power and status was threatened and who frequently had ideological objections, along with some peasants in well-run and efficient communes, who in a less doctrinaire manner were fearful that the breakup of the commune might harm their material interests. Indeed, the Third Plenum of December 1978 argued that 'it is forbidden to fix output quota or to distribute the land according to the individual households' (Riskin, 1987, p.286) and in March 1979, an editorial in the *Peoples' Daily* criticized the spontaneous rural reforms as going against historical trends, arguing that the three-level system used for production in the Peoples' Communes, with ownership by the production team as the basic form, 'could not be shaken' (China Pictorial Publishing Co.(ed.), 1989, p.17). But the grassroots movement was unstoppable and the government ceded its position. In September, 1980, Central Document no. 75 was issued from Beijing, which acknowledged the extensive debate on the responsibility system in rural areas and for the first time spelled out the circumstances in which individual household contracting *(bao chan dao hu)* might be allowed, notably in poor and backward areas where the population had lost faith in the collective and where household contracting had *already* been carried out and was working satisfactorily (Riskin, 1987, p.287). By 1982, household contracting (either *bao chan dao hu* or *bao gan dao hu)* was adopted formally by the central government as orthodoxy and enforced, even in those communes that were opposed to reform, thus reviving the Maoist tendency of imposing uniform policies across without heed for local views or conditions (White, 1993, p.105). As Kelliher writes,

> The final irony of family farming is that what began as an innovation by
> communities seeking more self-determination ended up being forced
> upon a minority of communities against their will. The state's original
> insistence upon uniform collectivization created a backlash of sufficient

coherence to convince the reformers to permit change. Then the same reform leadership reverted to the same uniform policy implementation - one cut of the knife *(yi dao qie)* as the Chinese call it - and forced the communities that preferred the collectives to give them up (1992, p. 106).

Bao chan dao hu allowed the household a particular amount of land which would then be contracted to produce output for the team in exchange for workpoints, any surplus production being kept for its own consumption or for independent marketing. Plans for planting, irrigation and the use of animals and machines remained under team control and the value of workpoints remained dependent upon team output; to that extent the income of each family was tied to the efforts of others. In the more radical *bao gan dao hu*, however, animals, machines and equipment as well as land was divided up amongst households. The authorities maintained a reserve planning power to allocate different activities amongst households, but essentially, after meeting its obligations to the state in the form of compulsory sales or taxes, the household was free to dispose of its output in any way it chose. For Riskin (1987, p.288), *bao gan dao hu* resembled 'tenant farming with the collective and state as landlord' (see also White, 1993, p.101).

Very quickly, the various forms of HRS spread through the countryside so that by mid-1982, 74 per cent of basic accounting units in China operated one form or another, by May 1983 this figure had grown to 93 per cent and in November 1983 to 98 per cent (Beijing Review, November 28, 1993, p.19). *Bao chan dao hu* peaked at 17 per cent in mid-1981 and thereafter declined in favour of *bao gan dao hu* (Riskin, 1987, p.290). Within four years of the Third Plenum, and despite the latter's initial opposition, Chinese agriculture was *de facto* decollectivised.

The rural reforms took forms other than the distribution of communal property and the implementation of the HRS. Free markets were encouraged and farm procurement prices were raised while at the same time the quotas for mandatory delivery to the state were frozen, allowing farmers to sell an increasing share of their annual output at the higher price paid for above-quota production. In 1979, the state increased the price of quota grain by 20 per cent while increasing it by 30-50 per cent for over-quota grain (Zhou, 1996a, p.5) and overall these two changes had the effect of increasing farm prices by 40 per cent in 1979-80 alone (Harding, 1987, p.102). And the restrictions on sidelines were abandoned, leading to a huge proliferation of non-farm activities and the development of households specializing in particular economic activities, such as duck raising or rabbit breeding. At the same time, the townships and villages sponsored an explosive growth in rural industrial enterprises. By the mid-1980s the political economy of the Chinese countryside had been transformed. In that:

> a vast array of economic decisions formerly in the hands of collective cadres was now taken individually by almost 200 million peasant households, the texture of daily life in the villages was transformed in a way unimaginable in the 1970s Nolan (1988, p.112).

33

3.3.2 The initial results of the reforms

In 1978, the grain crop amounted to 304 million tons, in 1980, 320 million tons, in 1984, 407 million tons (Chinese Statistical Yearbook, 1995, p.347). Such was the spectacular growth of the harvest in the early years of the 1980s that it is unsurprizing that the central authorities became, in Stalin's words 'dizzy with success' (White, 1993, p.108), believing that the reforms, by essentially giving head to the farmers' enthusiasms through the incentives provided by the reinstigation of family farming, had effectively solved the agricultural problem. Not only did the grain crop increase fast; so did the output of just about every product as households were able, once the grain contract was met, to use part of their land for growing high priced commercial crops best suited to local soil conditions. As one farmer explained to Unger,

> We were forced to grow rice even on swampy ground, with miserable
> results; today we raise water chestnuts there (1985, p.598).

And there is no question that in the years from 1980 onwards the average living standards of farmers leapt up. In 1978, the value of per capita consumption by rural residents was 138 yuan, by 1980 it was 153 yuan, and by 1985 it had jumped to 267 yuan in real terms (China Statistical Yearbook, 1995, p.258). The diet of the average rural household became significantly more varied with higher per capita consumption of meat up from 5.76 kilograms in 1978 to 10.97 kilograms in 1985, of poultry from 0.25 kilograms to 1.03 kilograms, of eggs from 0.8 kilograms to 2.05 kilograms and of fish from 0.84 kilograms to 1.64 kilograms (China Statistical Yearbook, 1995, p.287). Meanwhile the consumption of consumer durables rocketed between 1978 and 1985: In 1978 there were 30 bicycles, 19 sewing machines, 27 wristwatches and 17 radio sets per hundred households, by 1985 there were 81, 43, 126 and 54 respectively (China Statistical Yearbook, 1995, p.287). Perhaps most significant of all, the per capita living space, which was 8.1 square metres in 1978 had jumped to 14.7 square metres by 1985. And the fabric of that living space had changed, from the norm of mud and straw in 1978 to brick in 1985. The statistics show a rise of 67 per cent in real per capita peasant incomes between 1978 and the end of 1982 (Unger, 1985). Meanwhile, the number of enrolments in rural kindergartens and primary schools mushroomed in the early 1980s as did the number of clinics. Thus the 'little short of revolutionary' (Per Ronnas, 1994, p.233) increase in private income and welfare was accompanied by improvements in welfare at the social level.

But these startling improvements in living standards were not based primarily on increases in the grain and other harvests between 1978 and 1984, however impressive. Nor were they based exclusively on the increased opportunities for households to engage in side-line activities, however rich some became as a result. Rather, they were primarily dependent on the proceeds of the rapid *industrialization* experienced in the Chinese countryside in the early 1980s. This process had already begun in the 1970s but the reforms at the end of the decade led to its sensational acceleration.

Per Ronnas (1994, pp.229-230) argues that there were five particular features of the Chinese rural economy in the 1970s which were highly conducive to rapid industrialization. First, there was a huge supply of cheap labour which, because of the stringent household registration system *(hukou)*, was 'captive' in the countryside. Labour intensity in agriculture was high even by Asian standards, its marginal productivity low and the scope for substituting capital for labour and deploying it in more productive activities was immense. Secondly, the work-point system meant that the communes and brigades had very powerful means of generating investible capital by forced savings: since the industrial enterprises run at brigade or commune level compensated workers hired from a production team by the average value of the work-points earned by the team as a whole, and as the productivity of enterprise workers was normally significantly greater than that of farmers, the workers were not paid the full value of their labour and the difference was creamed off by the enterprise. Thirdly, the supply of inputs, such as locally produced raw materials, eased during the 1970s with increased autonomy from the centre and greater opportunities for making horizontal contacts; fourthly the inability of state planning to satisfy increasingly varied local demands provided ideal market conditions outside of the state network; and fifthly, the three-fold commune-brigade-production team administrative set-up provided good conditions for the development of a diversified local economy, the different tiers allowing activities to be established and run at appropriate levels of scale while simultaneously sharing risk.

However, after the reforms, the situation improved still further. While the reforms reduced the brigades' and communes' abilities to generate forced savings, the local enterprises benefited from increased local autonomy, from rapidly expanding (and largely captive) markets both for supplies and for outputs, as rural incomes raced upwards, while still enjoying the benefits of a captive labour market, since there was no official weakening of the *hukou* system, and the cosy three-tier institutional structure.

Thus on the basis of this very favourable political economy, rural enterprises, often referred to as township-and-village-enterprises [TVEs] *(xiang zhen qiye)*, mushroomed. In 1978, there were 1.52 m. such enterprises, employing 28.3 m. people and producing an output value of 49.3 b. yuan. In 1985, there were 12.2 m. enterprises, employing 69.8 m. people with an output value of 272.8 b. yuan (China Statistical Yearbook, 1995, p.365). Though there was some development in private ownership, the majority of enterprises in the early years were owned collectively and run by the cadres at commune and brigade level. Many erstwhile communes merely changed themselves into companies: one moment they were *gongshe,* the next *gongsi.* While heavy industry (steel, cement, chemical fertiliser, machinery and power) had dominated rural industrialization in the 1970s, increasingly the TVEs moved downstream into agri-based light industry in the 1980s as profits from this source soared (Per Ronnas, 1994, p.231).

3.3.3 The rural political eonomy since 1985

Perhaps the greatest apparent 'success' of the reform period has been the continued development of the TVEs. Indeed, the spectacular economic growth enjoyed by the Chinese economy as a whole has been primarily dependent upon continued growth in the rural industrial sector. In 1978, the gross output value of TVEs was 49.3 billion yuan while total industrial output across China was 423.7 billion yuan, representing 11.6 per cent of that total. By 1995, the gross output value of TVEs was 6891.5 b. yuan, with a total industrial output in China of 9189.4 b. yuan, representing 75 per cent of that total (China Statistical Yearbook, 1996, pp.389 and 401). As a result of these spectacular statistics, material living standards for the majority of rural dwellers have continued to rise unabated. In the early 1980s the slogan *'wenbao'* (meaning to have warm clothes and be able to eat one's fill) was replaced by the concept of *'xiaokang'* (small wealth ie. being 'comparatively well-off') as the desirable state-of-affairs by the year 2000 (see Liu, 1987, p.32). This concept is a progressive one in that it attempts to go beyond GNP statistics as indicators of development and to include other variables such as quantity and variety of diet, the percentage of income spent on food, better and more varied clothing, higher standards of education and improvements in cultural life (Delman, 1990, p.42) and it is clear that for the vast majority of Chinese, in the countryside as well as the town, *xiaokang* is already the felt experience. Poverty is still suffered by many people (a small percentage of 1.2 billion people still represents a lot of hungry mouths), but as the World Bank (1995b, p.6) admits, the incidence of poverty saw dramatic reductions in the 1980s and early 1990s.

3.3.4 Sustainable development?

Despite the general euphoria concerning the (obvious to the eye) material improvements in rural welfare in China enjoyed over the last twenty years, there is a major question-mark over their *sustainability*. And the question-mark refers to the ability of agriculture to sustain its fundamental role as the basis for growth in other sectors of the economy. Well before Lester Brown asked the question 'Who will feed China?' (1995), many Chinese and overseas scholars (see Hinton, 1990) were worried at the direction agriculture was taking. While Chinese leaders put on a brave face when discussing future agricultural trends, it is clear they remain worried about food security. Any complacency which the record harvests in the early eighties might have engendered has now been dissipated (Ash, 1992, p.546).

3.3.5 Recent trends in Chinese agriculture

In 1984, the reported grain harvest was a record 407.3 million tons, having risen from a little over 300 millions in 1978. In 1985, however, the harvest fell back to 379 million tons and did not regain 1984 levels until 1989. By 1995 the harvest was up to 466 millions, but the per capita grain output was still lower [0.38 tons] than in 1984 [0.39 tons] (China Statistical Yearbook, 1996, pp.69 & 371). In the short-term, these declining per capita harvests may not appear too serious in the context of huge increases in the production of other crops, including peanuts, tobacco, silkworm cocoons, fruit and vegetables (China Statistical Yearbook, 1996, pp.347-8) and in the output of meat and fish, but given the importance of grain, and specifically rice, to the basic diet of the Chinese (as well as the need to feed increasingly large numbers of animals) coupled with traditional government sensitivity to the size of the grain crop, it is hardly surprising that the performance of Chinese agriculture remains a cause for concern. The record 1996 grain harvest, at 504m. and the 1997 harvest of 494m. (China Statistical Yearbook, 1998, p.386) gives transient comfort.

In 1985, the Chinese government, flushed with the apparent early successes of the HRS introduced a series of new market-orientated rural reforms. Most importantly, the system of mandatory state procurement was abolished thus reducing the role of the state as a monopsonist and increasing the role of markets resulting in a 'two-track' system (White, 1993, p.108), combining state regulation through contracts with free markets. Other reforms encouraged farmers to move away from multi-crop subsistence and towards more specialized forms of cultivation and encouraged the pace of rural diversification as labour was displaced from the land. But these reforms were 'rooted in an over-optimistic view of the future of agriculture' (White, 1993, p.108-9) and the collapse of the harvest a year later brought the government up sharp.

That the rate of growth of the grain harvest should slow down in the second half of the 1980s was not surprising in that the initial high growth rates began from a relatively small base. Engel's Law ensures that as incomes rise, a smaller proportion of that income is spent on basic foods anyway. And the dramatic organisational changes which clearly had a major impact on farmers' incentives could not be indefinitely repeated. However, it was not only the organisational changes which encouraged the farmers to produce more: the increase of 40 per cent in effective farm procurement prices in 1979-80 had a predictable effect on output. Thus, the reforms of 1985, which cut farm procurement prices, had equally predictable implications. Indeed, by 1986, the payments to farmers were well below what they had been receiving in 1983: 10 per cent lower for grain and 13 per cent lower for oil crops and for cotton and in at least half the regions of China, the peasants revenue from their whole mix of crops fell (Kelliher, 1993, p.137). As Putterman (1993, p.5) notes,

> The state's resolve to improve the terms of trade between agriculture and industry in favor of peasants, which bears much responsibility for increased farm incomes and production between 1979 and 1984, had definite limits.

Additionally, state investment in agricultural capital construction *fell:* in the period 1976-80 agriculture obtained a 10.5 per cent share of total capital construction investment, in the period 1981-5, it fell to 5 per cent (Brugger and Reglar, 1994, pp.127-8) and continued to fall subsequently (White, 1993, p.109; Ash,1991, p.518). According to Yao (1994, p.61) lack of state investment was one of the two key factors in explaining the stagnation of grain production in the latter part of the 1980s, the other being the failure of price reform in grain (see Chapter 5).

Thus there were a number of obvious reasons why agricultural output failed to keep up the pace of the early 1980s in the second half of the decade. However, there were also a number of other, less obvious and more deep-seated socio-economic factors which cumulatively may put a question-mark over the social and ecological sustainability of agricultural production beyond the short-term.

4 Rural Reforms and the Sustainability of the Chinese Rural Political Economy

4.1 Decollectivization

Decollectivization has been greeted with very different reactions from scholars, many of whom (e.g. Unger, 1985, Ross, 1988, Lin, 1990, 1992) accept the breakup of Maoist collective structures as desirable, progressive and *inevitable* while others, (e.g. Chossudovsky, 1986, Hinton, 1990, Muldavin, 1992, 1996a, 1996b) paint the process in almost wholly negative terms. Chossudovsky, for example, sees the process as responsible for restoring the rich peasant economy, proletarianizing the poor peasantry, encouraging rural-urban migration (as the basis for the development of a free market in hired labour), reinforcing social inequality, downgrading rural social welfare programmes and partially privatizing health and educational programmes (1986, pp.59-76). Muldavin laments the mining of *communal capital* (1992, p.4).

While accepting that some achievements *were* made during the Maoist era, specifically citing the commune system as providing minimum standards of welfare when regimes in other countries generated large numbers of landless and destitute, Putterman (1993) argues that the generally favourable increases in yields were due primarily to a:

> sharp intensification of input application. Extension of irrigation, increasing use of chemical fertiliser and dissemination of improved seeds, the hallmarks of what has been known elsewhere as the 'Green Revolution' were successfully promoted, in part on the basis of indigenous technical innovation (1993, p.348).

Moreover, Putterman goes on to argue that the achievements of collective period would have been greater but for four factors: the central government policy of squeezing the agricultural sector in the interests of the industrial sector by deliberately loading the terms of trade against the countryside in the setting of farm procurement prices, the egalitarian and anti-incentivist character of the collective units, the coercive nature of the regime and the intrinsic diseconomies of the group form of farm organization (1993, pp. 349-50).

Nolan (1988, ch.2) discusses at length the economic, political, and social reasons why socialist societies have tended to favour collectivized agriculture: (i) because an economically independent peasantry might pose a political threat, (ii) because collectives may be able to raise the rate of rural saving and investment, (iii) because state marketing collectives can serve as vehicles through which the fruits thereof can be syphoned off into non-farm activities, (iv) because collectives may serve as vehicles for the rapid diffusion of new techniques, (v) to prevent class polarization, (vi) to enable basic needs to be met more effectively and (vii) to realize economies of scale and overcome problems of 'lumpy' investment. In each case Nolan argues that either the advantages are exaggerated or are realized better in some other institutional form and the disadvantages minimized or ignored. Indeed, Nolan (1988) argues that:

> there are few advantages and a great many disadvantages to full collective organisation of the rural economy, so that it is most unlikely that peasants would voluntarily adopt such an institution (p.34).

And

> it is surely cause for reflection that the two worst famines of the twentieth century (in the Soviet Union in the early 1930s and in China in 1959-62) have occurred in countries with collective agriculture (p.49).

And

> it seems unquestionable that a major part of the problems encountered by the rural economy pre-1978 is attributable to the collective farm *per se* (p.79).

Thus it is unsurprising that Nolan should conclude that while there were progressive changes in the macro-environment experienced by farmers in the early 1980s - improvements in marketing systems and the agricultural 'terms of trade' - it was the organizational changes at the micro-level which did most to improve the rural sector's performance (1988, p.113). This largely accords with the view of Perkins and Yusuf (1984, p. 73) when they note that:

> What sets China apart from other developing countries is not that it has increased the use of chemical fertilisers, but how much it has relied on organizational reform to achieve rural development (quoted in Lin, 1990, p.149).

Macmillan, Whalley and Zhu (1989), using econometric analysis come to conclusions supporting this analysis, viz. that 78 per cent of the improvements in Chinese agriculture between 1978 and 1984 was due to improvements in incentives (as a result of the introduction of the HRS) and only 22 per cent was due to increases in farm procurement prices (1989, p.781). Lin (1989, p.34) suggests that decollectivization improved total factor productivity to the extent of accounting for over 50 per cent of the output growth between 1978 and 1984. And Putterman (1993) accepts that much of the additional income earned by the peasants in Dahe Township in Hebei province, the site of his principal fieldwork, was due to the organizational reform process and not only to the higher prices and market incentives,

> The transfer of resources from lower to higher value activities... also played a part in peasant income gains. What made this transfer possible was the loosening of administrative controls over resource allocation and the greater efficiency in farming that resulted from better micro-organization and from the freedom to pursue other activities in the time saved by more intensive work (1993, p.351).

Yao (1994, p.7 and p.61) agrees by concluding that the HRS has been the key to the success of the rural reforms, which with greater market liberalization has expanded individual incentives while 'the abolition of the commune system has substantially reduced the overheads of agricultural production'. Meanwhile Unger (1988) would agree with Nolan that decollectivization had other beneficial political-economic effects:

> One of the most beneficial consequences of de-collectivization is that the arbitrary power of the state and rural officials to exact cowed compliance from the peasantry has been weakened (1988, p.133).

But to the extent that the performance of the agricultural sector in the late 1980s and early 1990s has not lived up to the promise of the early 1980s, particularly given the consolidation of the HRS and greater (though not complete) price liberalization since then, it may well be that other forces were at work which help to explain the rapid rise of the harvest in the early 1980s and its uneven development since.

4.2 After All, Who *Will* Feed China?

That uneven development has led to a shrill, yet critical debate over the future ability of Chinese agriculture to feed its population, a debate which began in earnest when Lester Brown publicly argued that, given what he considered to be a reasonable scenario of projected demand and supply conditions, by the year 2030, 'the import deficit would reach 369 million tons, nearly double current world grain exports (1995, p.97).' So sensitive is China to the problem of food security that Brown's work has come in for considerable criticism, most notably from the Chinese government itself. As Brown explains, the Chinese ambassador to Norway called a press conference *within twenty-four hours* of his making the first comments to the above effect (at an international conference in Oslo), claiming that his analysis was 'off-base' and 'misleading' (1995, p.16), arguing that China was 'giving priority to agricultural productivity' (p.17) and pointing out:

> unequivocally that China does not want to rely on others to feed its population and that it relies on itself to solve its own problems (p.17).

Since then, the pages of China Daily (viz. 'Chinese farmers can feed the future', by D. Gale Johnson, July 11 and 12, 1995) and Beijing Review (April 8, 1996) have been used to rubbish Brown's assumptions and thus discount his conclusions. In November, 1995, at the Fourth European Conference on Agricultural and Rural Development in China (ECARDC IV), Yao Jianfu from the Chinese Ministry of Agriculture took up the debate by blaming the statistics on which Brown and others have drawn their conclusions, arguing,

> Who will feed China? China itself! The real amount of the farmland and per area yield in China are much more than the figures published on the statistical yearbook. According to .....Professor Lu Liangshu, there are 1.3 billion hectares farmland in China, therefore China has a great potential in land productivity (1995, p.1).

It is not just the Chinese government which questions Brown's results. With different assumptions about the behaviour of key variables, Alexandratos (1995) comes to different, less spectacularly awful, results, as do Wang and Davies (1996) and Huang, Roselle and Rosegrant (Lin, Huang and Roselle,1996 pp.78-85), who attempt to account in their 1995 model for a wide range of factors, including, on the demand side, urbanization and market development and on the supply side, technological change, agricultural investment, *environmental trends* and institutional innovations. Of course, the debate highlights not merely the question-marks hanging over Chinese agriculture, but the difficulties (indeed, absurdity) of taking econometric modelling too seriously when so many variables are involved, when those variables are exceedingly difficult to predict on the basis of past experience and when very small differences in assumptions about their future behaviour produce very different results. And, indeed, we can add a further difficulty: the problem of relying on statistics which even Chinese Ministry of Education officials admit in international conferences to be wrong.

Thus, while the future state of Chinese agriculture is a topic of intense debate which cannot be resolved by simple econometric modelling, and while there are many who will defend China's ability to feed its people, it is important to ensure whether China *will* indeed be able to do so without too much difficulty into the foreseeable future. Yao (1994, p.7) suggests three reasons why it is important for China to achieve long-term food self-sufficiency: the comparative advantage that China possesses in food production, the need for stability of supply in domestic markets and the need to save precious foreign exchange. However, it is also important, because as the Chinese government has recognised since 1949, the agriculture sector must be sufficiently strong not just for its own sake but also for for the economy to be able to 'walk on two legs', to enjoy balanced, *sustainable* development. A commonly expressed sentiment at the grass-roots is 'without agriculture, no stability, without industry, no riches'*(wu neng bu wen, wu gong bu fen).*

4.3 Decollectivized Agriculture: the Question of Sustainability

Though the influence of the market has expanded since the reforms of the late 1970s and early 1980s as price controls have been progressively though unevenly abandoned, and while the fabric of daily life has been substantially transformed in terms of employment and consumption, the forms of landholding and the responsibility for and scale of farming in most parts of China have remained almost unchanged. In most villages, farming remains the responsibility of the family, the scale of its operations remains remarkably small and performed with high labour- and low capital-intensity. In 1995, the area of cultivable land per household averaged only 2.17 *mu* per capita, but was lower than 1 *mu* per capita in Fujian and Zhejiang provinces and Beijing Municipality and less than 1.5 *mu* per capita in another 11 provinces, mostly in the south and east (China Statistical Yearbook, 1996, p.343). The small-scale nature of farming was accentuated at decollectivization by the egalitarianism with which land was distributed, households often being allocated two or three plots of land of differing qualities and locations to ensure that each would have its fair share of good and bad land. The size of the resulting fields was thus further reduced: in many of the old collectives, the land was divided to resemble patchwork quilts and most families found themselves farming fields no bigger than 'noodle strips' (Hinton, 1990, p.14).

Meanwhile, in many parts of the Chinese countryside, particularly in the poorest provinces in central and west China, traditional wooden ploughs are still pulled by donkey or water buffalo when not pulled by hand. Across China as a whole, in 1994, while there were 588 draught animals and 407 handcarts with rubber tyres per 1000 rural households, there were only 88 mini and walking tractors, 79 large and medium tractors and a mere four motor vehicles (China Statistical Yearbook, 1995, p.342). Though this represents marginally higher capital intensity than in the 1940s, other characteristics, including individualization, scale and labour intensity, are uncannily the same as then, when Chinese agriculture was reckoned by most scholars to be hopelessly inefficient and in need of reform (see Chapter 3). In particular, the fragmentation of the land in the 1980s has led to all the problems that Riskin complains of when discussing the same process in the 1940s, wastage of land, labour time, equipment and animals with an increase in the difficulty of efficient irrigation and drainage and the number of inter-farmer disputes over access (Riskin, 1987, p.23), those same factors which led him to suggest that the extant 'socio-economic conditions in the villages ... hinder(ed) investment, innovation and growth' (1987,p.32).

Greater availability of rural savings in the 80s than in the 40s or 50s has allowed higher levels of net investment in the later period and made improvements in productivity in the early 1980s possible as a result. However, fragmentation of land has been exacerbated by absolute losses in the area under cultivation. China's statistics in this context are sobering, particularly given the often-quoted demands on Chinese agriculture to feed 22 per cent of the world's population on 7 per cent of the arable land (Edmonds, 1994, p.6, NEPAa, 1991, p.1, Cheng, Han & Taylor, 1992, p.1127). The total sown area in 1994 was less than the 1978 figure by a total of 1.8 million hectares, representing a fall of 1.25 per cent but the loss of land sown to grain was much larger: over 11 million hectares, representing a loss of 9.15 per cent. The loss of land sown to rice was still greater, at 12.34 per cent of the 1978 figure (China Statistical Yearbook, 1995, p.344). Thus in a period when the rural population *rose* by 65 million (more than the entire population of the UK) and when the nutritional expectations of the population at large (which rose by a massive 236 million) were ever more demanding, the land area to feed them diminished. Losses result from many causes, but clearly residential building both at the edges of towns and in the villages has been a major source of loss. Other sources include the building of transport links, industrial and mining enterprises, paths and ditches (Smil, 1993, p.57) and, indeed, the still officially discouraged practice of digging graves in the middle of fields has been revived, thus reducing the availability of arable land still further. But farmland loss is also the result of environmental degradation, in particular soil erosion and desertification (see Chapter 5 below, Smil, 1993, p.57).

As rural industry and hence employment prospects have blossomed in the countryside, farming across large sections of the Chinese countryside has now become a largely part-time occupation, often performed by children, the elderly and women. While in the mid-1990s families still farmed small plots of land, frequently maintaining some livestock in the form of a few chickens or a pig, the majority of household income is generated by family members, mostly young and male, labouring in village enterprises increasingly leaving the work in the fields to the elderly and to women. As Li (1995, p.9) observes, there has been an increasing feminization of Chinese agriculture:

> According to the statistical data of 1993... in some cotton-growing areas, tea growing areas and silkworm raising areas, more than 90% (of the) labour force are women.

At the end of the 1970s, almost all peasants, young and old, male and female worked on the fields collectively producing mostly grain and earning work-points, living in mud-and-straw huts, without running water, electricity or consumer durables, relying for cooking, heating and sleeping on the traditional *kang*. The 1990s painted a very different picture of rural life, however. In this picture, most of those in the fields would be women or elderly members of a household privately farming small plots of land still mostly by hand (but with increasing amounts of chemical fertiliser) to produce food for the family's own consumption, to fulfil contracts or to sell in local markets, while the other members of the household would be working (if they hadn't already migrated to the town) in local factories, mostly collectively owned, producing bricks, or padded jackets or whatever, earning the greater part of the growing family income, increasingly spent on modern consumer goods, including colour televisions, video-recorders, produced in other parts of China or abroad.

And it is the argument of this chapter that the above picture is *unsustainable*. It will be argued that the current political economy of the Chinese countryside, based as it is, despite the changes wrought in the last 15 years, on the foundation of agriculture, is becoming increasingly unsustainable as the low material rewards from agriculture not only reduces the willingness of farmers to engage in agriculture at all, undermining the *social* sustainability of agriculture, but where they continue to work in the fields, encourages them to use methods which increasingly undermine its *environmental sustainability*.

4.4 Social Unsustainability

Deng's agricultural reforms unleashed the enthusiasm of the farmers but it has been for making money rather than for adopting beneficial agricultural practices *per se*. Where the two have coincided, well and good, but where they haven't, the first impulse has taken priority, leading to a widespread exodus from and neglect of the land and/or the incentive to maximize short-term output with little concern for its long-run viability. The reforms gave to all households their own plots of land, yes, but plots so small that mechanized forms of production are frequently ruled out and the simplest of economies of scale unobtainable. Though economists are prone to exaggerate scale economies (Schumacher, 1973, pp.10-17), it is surely difficult to sustain an argument that the most productive or in any way ideal farm size is 0.145 hectares per capita of household, which is the average farm size across China as a whole. (Li, 1995,p.10).

43

As Li (1995, p.9) notes,

> Because of the large population and limited land resources, the
> household scale of crop farming is very small. The small scale of crop
> production has resulted in low productivity, especially in East China
> where the economic level is much higher than other regions.... The
> small farming scale also affects the use of agro-machinery, especially
> the large-scale machines.

And she adds (1995, p.7)

> low economic profit of crops farming is the most restrictive factor
> *blocking the sustainable development of crops production in China.* The
> reasons for low profits (include)...small production scale (emphasis
> added).

Hinton (1990, p.116), on the basis of his 50-odd years of direct experience of Chinese
agriculture, pulls no punches,

> The heart of the matter remains: mechanization, to be successful, must
> have some scale. To go all out at the 15 horsepower level is *to condemn
> China to a chronically backward agriculture* (emphasis added).

The negative implications of small-scale agriculture are disputed by some western scholars.
Aubert (1995) and Pennarz (1995), for example, take more salutary positions: For Aubert:

> Rather than authoritarian measures aiming at creating 'economies of
> scale' (usually by helping the establishment of a few big specialized
> farms at the expense of ordinary villagers), it seems that the problem (of
> agricultural development) can only be be solved progressively by the
> development of outside job opportunities, with the temporary expansion
> of part-time farming (1995, p. 55).

While Pennarz suggests that

> For *sustainable* solutions to agricultural development, the diversified
> land use systems of smallholders should receive more support in order
> to allow small-scale economic systems to become an economically
> viable alternative to the labour-division type of agriculture (1995, p.19).

Indeed,

> Small-holder land-use systems show ecological benefits since they
> substitute external energy inputs (as fertiliser, pesticides, mast) with
> human labour which is reproduced within the subsistence mode of
> production (1995, p.19).

It must be remembered that Aubert and Pennarz base their conclusions on research done in
relatively poor parts of the Chinese countryside. But there are many who are not happy with this
point of view if applied to the country as a whole. According to Shao (1992, p.20),

> A micro-economic mechanism for rural China must...be able to solve
> the problems of agricultural operations *on an appropriate scale...* A

44

central issue in China's agricultural transformation is the transfer of surplus labour, yet if labour is transferred and land is not concentrated, labour transfer will not serve agriculture itself (emphasis added).

Meanwhile Li Bingkun (1995. p.3), an official of the State Council used the pages of China Daily to urge China 'to encourage systems that facilitate *scaled* and efficient development of agriculture (emphasis added).'

There are, of course, many other reasons advanced for agriculture's low profitability. Li (1995, p.7) adds the high prices of agricultural inputs and the low educational level of the farmers, accentuated by the feminization of agriculture already referred to. A commonly stressed factor leading to the relatively poor grain performance in the post-reform period (e.g. Wehrfritz, 1995, p.9; Lin 1990, p.162; Zhou 1996b., pp. 2-4; Putterman 1993, p.353; Yao, 1994, p.61) is the continued government intervention in the market for grain and the low state procurement prices, which has magnified the gap between government and market prices after 1985 when market prices increased partly as a result of the smaller harvests, thus reducing incentives to fulfil quotas. According to Lin (1990, p.162),

> marketing of most crops has been liberated. Grain is among the exceptions. Farmers are still required to meet the grain quota obligations at government-set prices. In addition, local governments often impose blockades on grain markets. This measure reduces the price of grain in areas with a comparative advantage in producing grain in the post-reform period, which contrasted with a sizeable growth of agriculture as a whole, can be attributed mainly to the decline in profitability of grain compared to other crops.

Thus it is clear that continued government intervention in the pricing structure of grain is perceived by many to be a serious handicap to future agricultural development. Ash (1991, p.518) makes clear that state financial support for the agricultural sector fell markedly in the 'decade of reform' as a result of an absolute drop in the value of fixed capital investment from the central budget and the fall in agriculture's share of total expenditure on capital projects, leading to a marked deterioration of the position of the farm sector relative to other sectors. But whatever reasons are prioritized, there is clear evidence from all over China that agriculture, particularly cereal growing, is not the route to material prosperity. As Wehrfritz (1995, p.9) noted in 1995,

> Losses are starting to mount. Farmland is built over as factories sprout like weeds. Farmland is lost to erosion. Peasants switch crops or move to the cities... This much is certain; China's 800 million peasants are tired of hauling in enormous grain harvests.... The result: 110 million rural refugees have flocked to the cities since Deng took power and the equivalent of all the cropland of Sichuan province has fallen out of production. Last year, the cumulative neglect finally produced the predictable result: farmers turned in a disappointing harvest.

In the words of Xiao Xingji (Director, Organic Food Development Centre, Nanjing Institute of Environmental Sciences, 1996, private conversation):

If the farmers earn five kuai a day in the fields and twenty-five kuai in
the factories, they'll work in the factories.

The latter comment highlights the role of rural industry in competing for human resources
in the village and thus highlights the double-edged implications of industrialization for the rural
political economy. If agriculture is not a route to material prosperity, in the extant conditions of
rural China it is not sustainable in *social* terms.

4.5 Environmental Unsustainability

Meanwhile, the record absolute harvests still being recorded in the mid-1990s are increasingly
dependent upon techniques which undermine long-run *environmental* sustainability. The
disturbingly fast rate of growth in the application of chemical fertilisers and pesticides, begun in
the Maoist period of collective farming, has gone on unabated. While the period 1978-84 saw a
doubling in the application of chemical fertilisers, the period 1984-95 saw a further doubling,
from 17.4 million tons to 35.9 million tons (China Statistical Yearbook, 1996, p.361). Coupled
with much more extensive use of chemical pesticides and increases in the area of irrigated land,
up by 5 million hectares between 1988 and 1995, despite an overall reduction of 0.8 million
hectares in the total area of land cultivated, agriculture in China has become ever more dependent
- despite low levels of mechanization - on energy imported into the Chinese countryside.
According to Cheng, Han and Taylor (1992, p.1127)

> During 1965-88, the total industrial energy used in China's agriculture
> increased by over nine times. The dominant energy use, for
> manufacturing inorganic fertiliser, increased by nearly twelve times...in
> 1988, the (latter) accounted for 83 per cent of the total industrial energy
> used in China's agriculture.
> In planning for the future, there are real concerns about China's capacity
> to increase fossil fuel energy supplies at a pace adequate to sustain
> continued grain yield increases required for feeding the country's
> immense and increasingly well-to-do population.

The trends of the 1980s have been sustained in the 1990s. Between 1988 and 1995, the
irrigated area expanded by 11.1 per cent, the application of chemical fertilisers by 67.8 per cent
and the total electricity consumed in rural areas by 132.5 per cent (China Statistical Yearbook,
1996, p.361).

It is not only the fossil fuel constraint caused by chemical fertiliser use which poses
environmental problems. The primary environmental problem associated with the use of
chemical fertiliser, especially nitrogen, is the leaching of nitrates into groundwater and runoff
into streams and surface water (World Bank 1992b, p.67). Other potential ecological problems
include the eutrophication of lakes and the hardening or crusting of soil. Meanwhile, the
increased use of chemical fertilisers has been accompanied by a levelling-off in the use of
organic fertilisers. Plants' ability to use nutrients efficiently is reduced as a result, crop yields
decline and to maintain those yields, the application of chemical fertilisers has to be accelerated:

the textbook case of diminishing returns. As Cheng, Han and Taylor (1992, p.1130) argue,

> There is a real concern that a further erosion in the relative importance of organic fertilisers will (militate against) realization of the increased yields otherwise possible from intensive inorganic fertiliser use.

Muldavin (1996b, p.298) emphasises the destructive consequences of increased chemical fertiliser use:

> In village case studies in Henan Province, yields have stagnated and declined because of soil degradation due to overuse of chemical fertilisers. Peasant farmers complain about the 'soil-burning' results of long-term fertiliser overuse and misuse. The soils became harder, less friable and the available nutrients diminished despite large additions of chemical fertiliser because of the loss of soil structure and decline in overall quality.

The increased use of chemical pesticides also poses environmental problems such as toxicity to the humans directly involved in their application, toxic residues in water, soil and food, and the increased resistance of pests to pesticides. The first is serious enough. According to Karen Janz (energy consultant, 1996, private correspondence):

> China uses the highest amount of agro-chemicals in the world. When I stayed in a normal village in Hebei last year, the farmers told us that every year several female farmers die because of chemical pesticides application in cotton fields. The amount is so high that a simple reduction would not be enough.

Smith (1997, p.20) notes that China's Ministry of Agriculture reported that more than 100,000 people were poisoned by pesticides and fertilisers during 1992 and 1993 and that more than 14,000 of them died.

By reducing its profitability, it is clear that the small scale of family farming has had negative implications for the sustainability of Chinese agriculture. On the one hand, many farmers, offered better opportunities in the towns and enterprises, have simply neglected the land, while others, unable to increase the productivity of the land in any other way (e.g. through mechanization or scale economies) have had to resort to the increased application of chemical inputs to maximize short-run returns.

Small scale has also threatened communal biogas development (see Chapter 6) and successful integrated pest management. According to Glaeser,

> Although IPM and modernization co-exist, the return of family contract-farming is gradually eroding the institutional base for IPM because it can only be effectively applied over large areas. Although it is conceivable that households could apply some of the IPM measures (for example, monitoring) little of this has been reported. (1995, p.100)

But it is also clear that another aspect of the reforms has further encouraged farmers to take a short-term view, the short period over which the leases were granted. Initially, it was for one or

47

two years, although in 1984, this was extended to 15 years (in 1994, this was extended to thirty years). Since farmers have had no reason to suppose that their plots would remain their own to farm for long (and, indeed, every reason to believe, if recent experience was anything to go by, that 180 degree shifts in central government policy would take them away at any moment) they have had every incentive to maximise short-run gain without any regard to the long term. It is hardly surprising, in these circumstances, that the application of chemical fertilisers and pesticides has increased so fast.

But it has not only been the dramatic increase in chemical inputs which has threatened the environmental sustainability of the post-reform rural economy. The rural industrial enterprises have been and still are frequently highly polluting, villagers showing little hesitation in running chemical works or paper factories or brickworks if they can make a profit, whatever their impact on the local environment. Meanwhile, the break-up of the collectives frequently involved an equal distribution of all collective property, including trees which were, in most cases, immediately chopped down for firewood, construction or for sale, provoking, according to Pu Maosen (official, Simao EPB, 1995, private interview) 'a level of deforestation not seen since the "Great Leap Forward" thirty years ago'. Marginal land was also brought into play. Hinton (1990, p.21) summarizes in characteristically colourful vein,

> The reform unleashed ... a wholesale attack on an already much-abused and enervated environment, on mountain slopes, on trees, on water resources, on grasslands, on fishing grounds, on wildlife, on minerals underground, on anything that could be cut down, plowed up, pumped over, dug out, shot dead or carried away.

And the break-up of the collective has had a further negative effect on the environment: it has become that much more difficult for village leaders to mobilize large gangs of villagers to build dams, or ditches or engage in reforestation or repair roads, to build up any kind of 'communal capital' or indeed, to deal collectively with environmental damage from whatever source. Thus, for Muldavin (1996a, p. 229), the abandonment of the communes and the introduction of the HRS have had serious implications for environmental sustainability:

> Much of the recent rapid economic growth (in China) has been achieved by *mining ecological capital*. The ongoing privatization of natural resources is accompanied by the shedding of risk to the lowest levels, forcing decision-making towards ever-shorter time-horizons. The privatization of the social (welfare, risk, communal capital), this personalization of risk and welfare needs, has profound effects on nature (emphasis added).

The mining of ecological capital that Muldavin refers to above involves resource degradation from many sources, including the increased use of marginal lands, more intensive cropping patterns, the substantially increased use of destructive chemicals, a decline in agricultural infrastructural investments, rapid exploitation of extant assets as a result of industrialization, accompanying industrial pollution, and the rampant destruction of ecosystems as a result of industrial mining and the search for construction materials (1996a, pp.236-7). And it is not only Western scholars who take this point of view. As Han (1989, p.803) notes,

48

The great changes in Chinese agriculture during the past ten years have been accompanied by significant deterioration of the rural environment, *as a cost of rapid development of the rural economy*. Among the most important environmental effects are soil erosion, destruction of forest and grassland vegetation, decrease of land quality and soil fertility, pollution by urban-industrial wastes and agricultural chemicals and the depletion of water resources [emphasis added].

Unlike Muldavin and Hinton, Han is careful not to blame the reforms *per se* for these outcomes. To the extent that the rapid development of the rural economy took place in the way that it did resulted from the reforms, however, it is possible to infer that those reforms had a direct impact on the extent of the 'mining of ecological capital' and that the post-reform development of the rural economy has been dependent upon such mining. Muldavin (1996b, p.289) goes on to suggest that the overall gains made in production and productivity were achieved by the mining not only of ecological capital but *communal* capital. Using evidence from long-term fieldwork in Heilongjiang province he notes,

With decollectivization and privatization, collective structures have experienced a massive decline in capital available for investment while simultaneously being stripped of assets and authority. Rapid acceleration of local-level 'natural disasters' in the last few years is largely due to the delayed effects of this decrease in capital investment and its environmental and therefore production consequences (1996b, p.301).

Thus, for Muldavin, the mining of both ecological and communal capital have resulted from the reforms with all the attendant 'production consequences' and implications for the sustainability of the rural political economy.

Scholars such as White (1993) and Edmonds (1994, 1995) do not take quite such a hostile or pessimistic view of the reforms in relation to their environmental consequences. Nonetheless, White is aware that concern has been expressed in China that:

the rural infrastructure (particularly irrigation works) built up during the commune period was deteriorating because the collective institutions which had built and maintained it had been weakened by the reforms....Concern was also voiced about the ecological consequences of the new stress on household agriculture and market exchange (a decline in soil quality caused by *'mining'* the soil for short-term profit, invasion and destruction of forests and pollution by unregulated rural industries [emphasis added](1993, p.109).

Edmonds (1994, pp.180-1), while suggesting that market forces have helped to control pollution in state-run industries continues by noting that:

market forces have engendered other environmental problems, particularly in the rural areas and the worst pollution in the next couple of decades will continue to be in eastern rural areas where small-scale industries proliferate.

And Smil, given his childhood socialisation in a part of Eastern Europe prey to wildly propagandist official sentiment, understandably so hostile to the Stalinist attacks on the environment during the period of Maoist collectivization, and while recognizing the benefits of *baogan dao hu* in terms of productivity, incomes and food availability, is aware that its introduction 'has not generated an automatic cessation of China's land degradation practices' (1987, p.222).

The pessimistic picture painted above should not be read to suggest that land distribution and the introduction of the HRS have closed off *all* possibilities of land consolidation or any large scale economically beneficial projects. The Poverty Alleviation Bureau *(Fupin Ban),* for example, in combination with provincial governments and ministries has mobilised large-scale land reclamation in Gansu and elsewhere in furtherance of poverty relief. (Indeed the village studies in Chapter 8 confirm that a small-scale 'privatised' agriculture is not fixed in tablets of stone). But in any event the evidence suggests that it is difficult to accept the view of Ross (1988, p.1) that market-based reforms have been generally beneficial to the condition of the Chinese natural environment and that the deepening and widening of those reforms provide the answer to continuing environmental problems. Indeed, according to the Agricultural Investment Research Group for the Central Committee of the CPC, concluding on possible policies for agricultural protection in China, and ever mindful of ecological problems, argues:

> *Rather than emphasize the market and circulation,* as in most developed countries, China should stress production, the construction of basic agricultural facilities and science and technology in order to prevent or ameliorate disasters (A.I.R.G., 1997, p.97).

4.6 Conclusion

Yao (1994) suggests that Chinese agriculture faces severe constraints on its future growth of four kinds: resource and physical, technical, institutional and economic. He includes in the former the loss of arable land due both to land erosion and urban development, frequent droughts and waterlogging in poorly drained areas and despite his hostility to the old commune system which, he argues, 'burdened agriculture with overheads' (1994, p.61) comments that:

> Current reports indicate that parts of the irrigation and drainage system, vital to increasing yields and stabilising production, have deteriorated. This is mainly to do with inadequate state investment in rural infrastructure. Without significant improvement in the physical and environmental conditions in vulnerable areas, the land productivity of many marginal areas will remain low (1994, p.103).

With regard to technical constraints, Yao suggests China has developed a good technical base to increase and maintain yields, but that this has taken place as a result of high input levels particularly of chemical fertilisers, which are at levels twice as high as in India and well above the average levels for developing countries. He thus concludes by suggesting that further increases in yield will be difficult to achieve except at high marginal cost. Noting that wheat and certain rice varieties suffer from a lack of disease, pest and water-logging resistance, he suggests,

> This may be restricting yield response to the higher rates of fertilisers, especially nitrogen fertilisers, now being applied and may be a main cause of yield stagnation in wheat and rice....
> The use of manure fertiliser has been rapidly replaced by the use of chemical fertiliser, primarily due to the increasing opportunity cost of labour in collecting manures and other organic fertiliser. This shift of fertiliser use can be very damaging to soil fertility in the long term (1994, p.103).

Regarding institutional constraints, Yao highlights the ineffective administration of the system of supplying fertilisers and agro-chemical inputs as well as the poor extension services and slow development of new technology. Additionally, he suggests that a key institutional constraint is the very low level of mechanization resulting from small size of the average farm, the large number of small plots, the lack of suitable agricultural machinery and the lack of spare parts and fuel. He berates the inadequate provision for depreciation and replacement costs made by the 37,000 *collectively run* mechanization stations (emphasis added) and suggests that 1.2 million small tractors and 160,000 medium and large tractors need replacing (1994, p.106).

Finally, Yao discusses economic constraints which he suggests are primarily the result of central government policy biases against agriculture, citing reduced state investment and the problems for grain marketing firms caused by the present system of grain procurement, arguing for a significant increase in the ability and flexibility of state marketing policy (1994, p.108). In this he is in clear agreement with Zhou (1996b. p.2-4).

While it is not difficult to concur with Yao with regard to the above constraints, there nonetheless appears to be a logical inconsistency and a serious generalized omission in his analysis. On the one hand, he argues strongly that the organizational reforms, specifically the introduction of the HRS, was the major factor in boosting agricultural output and yields, judging the communes to have burdened agriculture with overheads as well as blunting initiative, and advocates increased liberalization and flexibility in marketing to solve the main 'economic constraint'. Yet he seems unwilling to accept that many of the other important constraints that he classifies as physical and resource (e.g. loss of arable land, soil erosion, drought and waterlogging caused by deteriorating irrigation), technical (e.g. high marginal costs due to inappropriate and overuse of chemical fertilisers) and institutional (e.g. small scale of farming, low levels of mechanization and low levels of state investment in agriculture) are all in various ways related to the reform process and the forms of agriculture it has engendered. Indeed, while Yao clearly approaches the reforms and their impact on the Chinese rural economy within a very different paradigm from Muldavin (1992, 1996a, 1996b), it would be difficult not to argue that, cumulatively, the constraints he identifies result from the 'mining of ecological and communal capital', as Muldavin suggests, and that to find solutions to them it is necessary to move beyond the mantra of increasing market flexibility.

What is additionally surprising about Yao's work is how the developments of and prospects for the future of agriculture and indeed for the rural political economy generally are discussed at such length in abstract from a generalized discussion of an environmental constraint. Though not explicitly stated, and mindful that he does mention soil erosion as a problem, Yao constructs a model of Chinese rural development which treats the natural environment as though it were an exogenous variable. To the extent that all of the constraints are treated ultimately as technical, they are judged amenable to - individual- technical solutions. It is the argument of this work that this is not the case.

Thus, while the productivity of the land has increased since the reforms as grain output has risen despite declining cultivated acreage, it has occurred as a result of a steep increase in both the application of ecologically damaging chemicals and in the energy intensity of agriculture in a country chronically short of environmentally 'clean' energy resources and in a rural environment increasingly degraded and polluted by industrialization, its arable land decimated by the spread of non-agricultural activities and its communal resources available for environmentally beneficial practices depleted. This situation is increasingly clear to Chinese scholars. Referring to post-reform agriculture, Tian argues (1997, p.35),

> Since the area of farmland per capita is small, intensive farming methods and the increased quantities of fertiliser, especially chemical fertiliser, are used. This results in the degeneration of oil and contamination of soil and water. A shortage of supply leads to over-felling of trees, over-grazing of pasture land and excessive use of ground water, all of which have a grave effect on the environment.

It will be made clear in the next chapter that serious environmental problems existed in the Chinese countryside well before the reforms took place and that extant environmental problems can *by no means* be blamed entirely on them. However, it is my argument that the reforms have had specific impacts that have put the sustainability of the recent developments in the Chinese countryside into doubt. It is in the light of this potential unsustainability, with all its attendant impacts on food security, political stability and the health of the industrial sector that the Chinese government has been increasingly drawn to the promotion of environmentally more friendly practices in the countryside such as ecological agriculture.

5 The Peoples Republic of China: the Environment and Environmental Protection

5.1 The State of the Chinese Environment

Qu Geping makes it clear that environmental damage was well advanced by the early 1980s. Of the period of the Great Leap Forward, he notes,

> Under a zealous drive, mineral resources were... exploited, resulting in startling losses and destruction to both topography and landscape. Biological resources, forests in particular, were seriously damaged, causing several losses to out eco-system. There was extensive destruction of the natural environment of our country.
> Some of the damages to the ecosystem caused during the 'Great Leap Forward' have not been totally regained even after more than twenty years. (Qu, 1991, p.212).

Meanwhile,

> During the 'Cultural Revolution' the country's environmental conditions deteriorated rapidly to an extent from which it was hard to recover (Qu, 1991, p.218).

Qu goes on to recount a litany of environmental disasters, blamed not unnaturally enough, given the need for political expediency even in the 1990s, on Lin Biao and the 'Gang of Four' who, when faced with 'great social concern' as a result of 'the spreading pollution and damages, turned a deaf ear to the common cry' (1991, p.212). Thus when Vaclav Smil published *The Bad Earth* in 1984, the extent of the environmental pollution and degradation which he catalogued was a surprise only to outside observers who had been fed on official propaganda or western apologetic (see Smil, 1984, p. xi and p.10, Smil 1993, p.xvii). And since 1984, a slow, but steady stream of literature both by Chinese and Western scholars has been published, all giving testimony, albeit with different levels of emphasis to a serious, potentially critical problem for China in relation to the state of its natural environment.

In 1984 Smil reported that China suffered from just about every environmental disaster available and concluded, hardly surprisingly, on a pessimistic note,

> The magnitude of China's accumulated environmental problems owing to the legacy of ancient neglect and recent destruction is depressing. The dimensions of the future tasks in population control, food and energy supply, and overall societal modernization are overwhelming, and the potential for further accelerated environmental degradation is quite considerable (1984, p.198).

By 1989, Smil was hardly more optimistic. Commenting on the events in Tiananmen Square of early June 1989, he suggests,

> there can be little doubt that, in the coming years, the particulars of socio-political arrangements will be less important in determining China's fate than the country's treatment of its badly deteriorating environment (1989, p.277).

And, in 1993, although noting that environmental deterioration has not gone beyond the point of no return, he concludes:

> China's deteriorating environment is already a major retarding factor in the country's quest for modernization. The prospect for the 1990s, indeed at least for the next generation, is for further decline in China's environmental quality, a decline that may find global repercussions in fostering the country's instability and turning it into the planet's leading environmental transgressor (1993, p.201).

Glaeser (1987) is frequently just as pessimistic about the state of the Chinese environment although somewhat elliptically suggests there may be possibilities for other developing countries of 'learning from China'. However, given that the few Chinese successes that he alludes to include biogas digestion, which, as will be suggested later, has had mixed fortunes at best, it may well be inferred from his work that the most useful lesson other developing countries should learn from China is *not* to follow its example. Indeed, Glaeser (1995) seems to accept as much when he notes,

> environmental policy was never fully or successfully implemented, despite its serious intention, consistency and bureaucratic support. China....simply lacks the financial resources...there was little incentive...there was little enforcement. (1995, p.106).

As a result, Glaeser agrees with Boxer (1991, p.290) that:

> environmental problems in China seriously threaten 1990s economic modernization plans.

Arguably the most optimistic observer of the Chinese environmental state-of-affairs is Ross (1988) who, while recognizing the baleful state of Chinese environmental degradation, is firmly of the opinion that:

> markets have a considerable potential to enhance the environment that has only recently been recognized by China's leaders (1988, p.1).

To the extent that his favoured policy prescriptions run with the vein of current Chinese political practice, at least with regard to the handling of the economy, then optimistic outcomes are not discounted by him. He Baochan (1991) writing from within China could hardly be more pessimistic, however. Though He's argument purports to be about China's environmental crisis, it reads more like a frustrated polemic against the Chinese political regime in general and its handling of the Chinese political economy since liberation in particular. The frustration is evident, for example, when the regime is blamed both for too much and too little interference in and concern for the people's welfare. Thus he is able to argue that:

> China's own mistakes ...are primarily the result of regarding the economy as a wartime system. From the 1950s to the 1970s, our basic development strategy was to strengthen centralized management....The key concepts underlying economic policy were guaranteed supply, production, allocation, adjustment, subsidies and so on (1991, p.43).

Yet at the same time, while commenting on imbalances in the economy, He argues,

> For many years, we have been harping on about the need to cut back capital construction projects, but in many places the result is like kneading bread: the more you squash it down, the more it expands. The state calls for concentrating capital to guarantee infrastructures like energy, transportation and other key construction projects, but in fact people everywhere are producing refrigerators, television sets, radios and washing machines and small-scale tobacco factories, fertiliser plants and breweries are springing up all over the place (1991, p.51).

As a result, He ends up with a confused prescription for the grim environmental picture he paints, both hostile to the centralized and autocratic planning of the state, yet mindful of the need for new laws and tough implementation of them, hostile to the forms and pace of Chinese economic development giving rise to environmental damage yet critical of China's low level of economic development and thus of its capacities to do much about it.

Other major contributions to the literature on the state of the Chinese environment have come from Edmonds and Vermeer. Both view the situation as grave. For Edmonds (1994, p. 156), 'China experiences virtually every type of environmental degradation and the situation is worsening', while Vermeer (1990, p.34) quotes approvingly a report from an international environmental conference in Beijing that 'nobody, least of all the Chinese officials involved denies that China is on the brink of an ecological catastrophe'. Meanwhile very recently, Smith (1997, p.19) notes that:

> no country in history has undertaken an economic and industrial revolution on an ecological foundation in such a degraded state.

There is thus no dispute nowadays both within China and without, that China's natural environment is in a mess.

5.2 China's Environmental Problems: a Review

It is clear from the above overview that all the major contributions to the literature over the past decade have painted a thoroughly depressing picture of the Chinese environment. As a backdrop, if not a cause of the problems, is the huge and still rapidly rising population. When Ma Yinchu, President of Beijing University, first published his concerns about the size and growth of the population in 1957, advocating family planning to deal with it (and being sacked from his job by Mao for his pains), China's population stood at 646 millions (China Statistical Yearbook, 1995, p.59). Forty years later, it has doubled, standing at a little over 1.2 billion in 1996. Altogether, nearly 23 per cent of the world's population inhabit China, a statistic which will continue to rise until the moderation of the currently high birth rates and low death rates [in 1995, 16.57 per thousand and 6.51 per thousand respectively, (China Statistical Yearbook, 1998, p.107)], the results of an extant population being highly skewed towards youthful middle age.

Though China is the third largest country by area in the world (after Russia and Canada), much of its land, particularly on its north-western and south-western borders is largely uninhabitable through climate or altitude, and there is thus severe pressure of population on the available natural resources. According to official Chinese sources (NEPA, 1992, p.11) China's land mass comprises 33 per cent mountainous areas, 26 per cent plateaux, 19 per cent basins, 12 per cent plains and 10 per cent lilly land: thus nearly two-thirds of China's land area is 'rugged' (NEPA, 1992, p.12) and the majority of the population - three-quarters - are concentrated in the north, north-east and central-southern parts of China, representing only 44 per cent of the land area (NEPA, 1992, p.15). The Chinese authorities (NEPA, 1992, p.12) suggest that of the total utilized land, farmland covers 96 million hectares, amounting to only 0.085 hectares per person. The availability of per capita habitable land, therefore, is low and of arable land still lower, only one-tenth of the amount available per head in the former USSR (Cole, 1994, p.31). Meanwhile, China's per capita level of land, forest cover and water resources are but 36 per cent,, 13 per cent and 25 per cent respectively of the estimated global per capita levels (Wang et al., 1989, p.1, quoted in Edmonds, 1994, p.156).

. Though clearly not the *cause* of environmental degradation and pollution *per se,* the high intensity of land-use which has resulted from the large and still growing population has contributed to the difficulties China has experienced in its recent quest for development and has made development that much more prone to rapid environmental damage.

Environmentalists will often prioritize deforestation as a cause of environmental degradation because of the wide-ranging negative impacts it can have. While China claims to have 'implemented massive ecological afforestation engineering projects (NEPA, 1992, p.23) since 1978, China's forest cover, at a little below 13 per cent in 1991 is well below the world average of 31 per cent (Edmonds, 1994, p.158) and the situation is handicapped, despite the official afforestation programmes and enactment of legislation (viz. the Basic Forest Law in 1985) by 'massive illegal logging', expanded since the arrival of the market economy (Edmonds, 1994, p.159), 'enormous fraud and waste' (Smil, 1993, p.59) 'inadequate prevention of forest fires and poor fire-fighting facilities' (Smil, 1993 p.62) and 'false reporting, ineptitude... bad management, theft and blatant looting' (He, 1991, p.27). According to Smil (1992, p.435) during the 1980s the forest areas decreased by 12. 8 million hectares, i.e. by about 10 per cent, while the excessive felling of recent years means that by the year 2000 nearly 70 per cent of China's forestry bases will have no trees to fell.

One of the many negative impacts of deforestation is soil erosion, claimed by Han to be 'the most critical problem in rural development' (1989, p.804). Soil erosion is a common and natural process but becomes serious when it is not offset by various remedial strategies, leading to decline in agricultural productivity and ultimately, if not checked, to advanced degradation and loss of the power of eco-system restoration (Smil, 1993, p.10). According to Edmonds, 'the major causes of China's soil erosion are deforestation, cultivation on slopes, overgrazing and poorly managed industrial land-use' and 'in most cases humans are responsible for (its) severity' (1995, p.63). Thus not only is agricultural productivity threatened, but cultivable land is lost in the process. Smil notes that the organic matter content of intensively cultivated soils in the North-East China plain fell from its natural level of 9 per cent to 2 per cent by the mid-1980s (1993, p. 58) and quotes a nationwide survey in 1988 of 45 million hectares in 901 counties across China which discovered various degrees of excessive soil erosion on 31 per cent of the land (1993, p.57). But the consequences of soil erosion in China are not limited to loss of soil productivity and land: additionally, it has increased the potential severity of both flooding and drought. Enormous losses of reservoir storage capacity are lost through siltation [e.g. in Shanxi], irrigation channels and tributaries blocked up [e.g. on the Loess Plateau] and river beds raised, in some cases above the surrounding flood plains [e.g. in western Shandong province]. (Edmonds, 1995, p.69-70). It is thus not surprising in a country sometimes plagued by drought, that serious floods, causing loss of life and damage to property, as in 1991 1996 and 1998, frequently occur. Thus the prevalence of so-called 'natural' disasters -floods, drought, forest fires- can frequently be explained by human action and the poverty that is often the human condition. As Edmonds notes,

> The cost of soil erosion to the Chinese people cannot be calculated in monetary terms. The relationship with poverty is clear, although how much soil erosion is a cause of poverty and how much poverty is a cause of soil erosion is not so clear. In any event it is said that of the 270-odd poorest counties in China, 87 per cent are in areas suffering serious soil erosion (1995, p.70).

China also suffers from serious desertification and according to Edmonds (1995, p.103-108) close to 2 per cent of China's total land area can be considered desertified *by human-induced* resource degradation, largely through overcultivation and overgrazing, mining, industrialization and urbanization, with an increase in the area desertified by about 21,000 sq. km. between 1975 and 1986, mostly in the grasslands in the north, northeast and northwest China. He (1991, p.21) suggests that already one-seventh of China's land mass is now desert and that without effective controls another 70.000 sq. km. will go the same way by 2000 (p.23). Though the latter numbers may be exaggerated, the threat that desertification poses, particularly though not exclusively in the grasslands, is clearly a substantial one, as is salinization-alkalization, the process whereby salinized or alkalized soil is produced through the accumulation of salts in the soil (Edmonds, 1994, p.124), primarily as a result of over-irrigation, which adversely affects 7.656 million hectares (China Statistical Yearbook, 1995, p.338) roughly one-sixth of China's irrigated cropland.

Smil (1993, p.37) reminds us that the efficiency of photosynthesis is determined by its four indispensable natural preconditions - solar radiation, atmospheric carbon dioxide, land and water. While there is no present or indeed future threat to the availability of the first two, the latter two give serious cause for concern. For not only is the quantity and quality of the land under threat, as we have seen, but China faces serious water shortages as well. Of China's 500 biggest cities, 300 are considered short of water, and 108 'acutely short' (Smith, 1997, p.19). And once again, these shortages, mainly on the North China Plain, are largely the result of human activity, including the extensive construction of reservoirs on the upper streams of rivers, intensive irrigation, the drilling of wells for pumping deep water and drainage for the amelioration of salinized lowlands and the increasing water usage of urban-industrial development (Han, 1989, p.206, Edmonds 1995, p.114-5). As an illustration of the impact of the latter, it has been suggested that, should groundwater levels continue to drop at the current alarming rates, China's capital, Beijing, may have to move by the year 2005 (Guardian, March 9, 1990, quoted in Vermeer, 1990, p.36; Smil, 1993, p.43: Smil, 1992, p.434). In the 1950s, the water table beneath the capital was only 5 metres below the surface, today it is around 50 metres below (Smil, 1993, p.43). And while urbanization, industrialization and irrigation will continue to put a heavy burden on water resources, the increased popularity of beer drinking and output of the brewing industry will only make things worse. All sorts of possible solutions have been tried, including large scale water transfer, but without conservation and economies resulting from greater efficiency in the use of water, on the land, in the factory and in the household, acute water shortage will continue to dog China's future development.

And stemming largely from the low levels and poor quality of waste water treatment, what water China does have frequently suffers from heavy pollution. As reported by Edmonds (1994, pp.165-6), studies from the mid-1980s suggest a litany of problems: that about 25 per cent of China's fresh water is polluted, with the proportion of lakes and rivers near big industrial cities polluted alot higher, that a survey of 95 rivers undertaken in 1990 suggested that 65 were polluted, that pollution levels in rivers could double between 1990 and 2000, that 80 per cent of urban surface water is polluted, that only 6 of 27-seven major cities can provide drinking water to standards set by the state, that 25 per cent of lakes are seriously or moderately polluted, that many lakes suffering from eutrophication have not supported fish for years and that that the number of such lakes is increasing and that there have been frequent alarming cases of marine pollution. Rural rivers are now suffering almost as badly as those in urban areas, as a result both of industrialization in the countryside, electroplating being the main culprit, and the increase in chemical fertiliser use.

Statistics for air pollution are equally sobering, hardly surprising, according to Smil,

> in a nation where combustion of one billion tons of coal, largely uncleaned and burned with minimal or no air pollution controls, supplies three-quarters of all primary energy (1993, p.117).

A particular problem associated with Chinese coal is its unusually high sulphur content (some Sichuan coals can contain up to 10 per cent sulphur content, and it is thus alarming that in 1990, less than 10 per cent of China's high sulphur coals are treated before combustion (Smil, 1993, p.118). Just about all forms of air pollution, including suspended particulate matter, SO_2, and NO_x, the latter two giving rise to severe acid rain problems, exist at levels well beyond even China's minimal standards. As a result, buildings, vegetation including trees, lakes and human health suffers (even the *representation* of colour suffers: the concept of 'greenness' is difficult to recognise or even imagine in many areas of China) while the financial costs that *directly* result have a major and immediate impact on the *welfare* of the Chinese. An under-discussed statistic is that life expectancy at birth has *fallen* in the last six years, *reducing* China's rank in the UNDP Human Development Index (UNDP Human Development Reports, 1990 and 1996), despite huge increases in GNP per head. Air pollution is likely to be a major cause of this depressing state-of-affairs, not least as a result of the air dragged directly into the lungs by China's ever increasing numbers of smokers of high tar cigarettes. But ever higher numbers of automobiles on the road, increasing presently at startling rates, will add further to the problem. Air pollution is already sufficiently so serious even in many of the major tourist centres, Beijing, Xian and Datong to name but three visited by the author recently, that significant indirect opportunity costs are also clearly involved, and the problem is getting visibly worse by the day.

Serious noise and solid waste pollution have developed only in recent times. Noise stems primarily from factories but increasingly from cars and the radios and television sets of neighbours. It is thus an obvious but unwelcome and costly (Vermeer, 1995, p.29) companion to the higher levels of consumption being enjoyed by today's Chinese. Meanwhile, solid waste disposal problems can be even more unwelcome. As Edmonds notes, (1995, p.150),

> Improper disposal of urban solid wastes can lead to to air and water pollution and the spread of diseases. The sight and smell of dumps can have negative psychological effects on neighbouring communities and affect the migrating habits of some animals.

The alarming rate of rubbish growth is primarily the result of the pace and forms of economic growth in China. But the increasing problems associated with the disposal of human and animal faeces is likewise, as recycling of wastes both from the town to the countryside and within the countryside has been increasingly reduced to being a low-status, messy business, which is happily avoidable for the recyclers, as a result of the availability of other income sources and for the recyclees, as a result of the availability of chemical fertilisers. Young girls carrying shoulder poles with big pots of excrement attached to each end are a much less common sight (and smell) today than only a few years ago. Soil pollution, partly the result of the increasing chemical fertiliser use and partly the result of industrial waste threatens to be an increasing menace, as has been the increased use of organochlorine pesticides in the countryside (Edmonds, 1994, p.168).

China is the home to a huge variety of flora, fauna and wildlife, the existence of which has come under severe threat in recent years. The regular loss of species is not merely a loss for China, but a loss for the global society. As Edmonds (1995, p.194) notes,

> the loss of one species not only reduces the world genetic pool in the loss of potential medicinal and industrial chemicals, food types and wild strains for genetic upgrading of cultigens, it can also trigger the death of other dependent species and set off a chain reaction of extinction.

The loss of biodiversity is merely *one* of the many global environmental implications of China's recent political-economic change. In that in 1995 the burning of fossil fuels accounted for 94 per cent of China's energy generation (74.3 per cent coal, 17.4 per cent crude oil and 2.3 per cent natural gas, China Statistical Yearbook, 1998, p.251) - and that it even obtains some energy from biogas, most of which is methane - China is a growing producer of 'greenhouse gases'. Moreover since China is one of the largest manufacturers of refrigerators in the world, it produces many of the CFCs responsible for the thinning of the stratospheric ozone layer. It can be convincingly argued, therefore, that it is not merely in the interests of the Chinese that the environmental problems of China should be addressed as a matter of the utmost urgency (Smith, 1997, p.25).

5.3 The Development of Environment Protection in PR China

The initial policies of reconstruction in the aftermath of the anti-Japanese and civil wars prioritized improving the lot of the urban population, ravaged by industrial destruction, criminality and hunger. As a result, the concept of 'environmental hygiene', borrowed from the Soviet Union, was developed and the provision of clean water promoted (Glaeser, 1990, p.249). By the time the First Five Year Plan was introduced in 1953, the environmental impact of industrial production (though not specifically mentioned) was recognised to the extent of it advocating the sparing use of natural resources and the recycling of industrial waste-water. But environmental problems were nonetheless manifest in the mid-1950s, particularly air and water pollution and in 1956 new legislation was enacted in order to improve the health of the urban population by ruling that industrial facilities should not be sited upstream and calling for the promotion of emission-abatement technologies (Glaeser, 1990, p.250).

Legislation barely touched the rural environment in the first twenty or so years after liberation although the enforced collectivization made it easier to mobilize gangs of peasants to construct or defend 'public goods' such as dams, reservoirs and irrigation schemes which had positive environmental impacts. Unfortunately, it was also easier to mobilize the same gangs to chop down trees to provide firewood for stoves and backyard blast furnaces as happened during the 'Great Leap Forward', with all the attendant environmental costs discussed earlier. Until the 1960s, few institutional or technical measures were taken to deal with burgeoning health and environmental impacts of industry, agriculture, urbanization and energy development (Boxer, 1991, p.294).

But by the early 1970s, even with the Cultural Revolution still ongoing, there was a noticeable change in attitudes toward the environment (perhaps influenced by the 'Limits to Growth' debate in the west), ecology became a political topic and the concept of 'environmental hygiene' was replaced with the concept of 'environmental protection' (Glaeser, 1990, p.251). That China was a participant at the first United Nations Conference on the Human Environment in Stockholm in 1972 was illustration of this turn of mood and with the advocacy of Premier Zhou Enlai (Qu, 1991, p.215, Boxer, 1991, p.294) used that conference as the springboard for the development of its domestic environmental policy, convening the first National

Environmental Conference in 1973 in Beijing. At that conference, guidelines were drawn up on environmental policy and environmental protection and research units established at different administrative layers for the first time (Glaeser, 1990, p.253). A year later the Environmental Protection Leading Group of the State Council was formed and in 1979 China enacted the first Environmental Protection Law of the PRC, thus marking the beginning of a decade in which both regulatory programmes and institutions for environmental protection expanded at an impressive pace (Sinkule and Ortolano, 1995, p.1).

In 1982 Article 26 of the Constitution - 'the state protects and improves the environment in which people live and the ecological environment. It prevents and controls pollution and other public hazards' (Constitution of PRC, 1987, p.22) - was enacted and in 1983 the then Vice-Premier Li Peng used the second National Conference on the Environment to declare environmental protection, along with family planning, to be a 'fundamental state policy'. According to Qu Geping (1991, p.226), the period from 1981 to 1987 was a 'golden age for development of environmental protection' with a panoply of new laws and regulations, further empowering environmental protection and environmental research units. In 1982, the state set up the Ministry of Urban and Rural Construction and Environmental Protection with the Environmental Protection Administration beneath it; in 1987 the latter was elevated from being an office of the Ministry to an agency directly under the State Council, giving it Cabinet level status, while its name was changed to the National Environmental Protection Agency, with Qu Geping as its chief, and given ultimate responsibility for overall supervision and administration of all environmental protection work in China. In 1989, this period culminated with the promulgation of the final and a comprehensive version of the Environmental Protection Law, exactly ten years after its initial, trial implementation. As Jahiel (1997, p.81) notes,

> The first decade of reform ushered in a vast expansion of environmental protection institutions, laws and policies, a process which in certain respects continues to this day.

In the 1990s, the status of environmental protection has been further consolidated and strengthened. In 1992, China took a full part in the UN Conference on Environment and Development ('the Earth Summit') in Rio de Janiero, with Premier Li Peng leading the delegation, and in 1993 a special Environment and Resources Protection Committee of the National People's Congress was established (State Council, 1996, p.10), headed yet again by Qu Geping, the 'father of Chinese environmental Policy', (Glaiser, 1990, p.253) while State Councillor, Song Jian, was given a formal environmental protection portfolio for the first time. In 1994, the Chinese government promulgated China's Agenda 21 - its response to the 'Earth Summit' and in 1996 adopted in the Ninth Five Year Plan firm commitments to 'sustainable development as an important strategy for modernization' (State Council, 1996, p.6). At the 15th Party Congress in 1997 the National Environmental; Protection Agency (NEPA) was strengthened, renamed the State Environmental Protection Administration (SEPA) and given ful ministerial status.

By 1996, China had set up a far-ranging network of 2,500 environmental agencies, bureaux and offices at a variety of administrative levels,from province, through city, county and township, employing a total staff of 88,000 people, and had drafted an array of (at least on paper) tough regulations concerning environmental protection, with infringement of environment legislation at its most serious punishable by death. But the response of the government, however sincere, has not matched the scale of the problem.

5.4 Future Issues and Constraints

Of the Western scholars of the Chinese environment and environmental protection, Vermeer (1990, 1995) is more optimistic than some. He is nonetheless aware that China faces formidable problems. Some of these are similar to other developing countries, he suggests: rapid industrial growth, low levels of productivity and income, a lack of specialist know-how, a low level of education, inadequate administration and financing, rapid population growth and loss of scarce, sometimes irreplaceable natural resources (1990, pp.35-6), Other problems he suggests stem from its unique, socialist, political-economic order, however: in particular, he cites the fundamental conflict between the interests of government as owner and operator of factories and its role as protector of the environment and public health, the undeveloped nature of the Chinese legal system and the lack of control exercised by local government over TVEs (1990, pp.36-7). Still more stem from the uniqueness of the Chinese physical geography: in particular, its unusually high dependence on unusually high sulphur coal for energy generation, its limited water resources and its sheer size (1990, p.38).

Vermeer (1995, pp.25-27) presents a numbers of Chinese studies of the economic costs of environmental pollution losses between 1980 and 1990, which vary from [at constant 1980 prices] 44 billion yuan (1980) to 88 billion yuan (1985) to 38.2 billion yuan (1990). Whether the figure is the highest (representing 12 per cent of GNP), the lowest (6.75 per cent of GNP) or something in between, the sheer scale of the numbers have persuaded China's leaders of the severity of the problem and the need to address it. In the 1980s, the government progressively increased the outlay on environmental protection from 0.4 per cent to 0.8 per cent and the absolute level has continued to rise to 14.5 billion yuan in 1993, although this represented only 0.5 per cent of an expanded GNP when NEPA experts have been calling for a minimum of 1 per cent and a desirable 1.5 per cent of GNP (Vermeer, 1995,p.32). The goal for the Eighth Five Year Plan (1991-95) was set at 0.85 per cent of GNP (Vermeer, 1995, p.35) and for the Ninth Five year Plan at least 1.0 per centf GNP (NEPA, 1997, p.15,NEPA/SPC, 1994,p.22).

Environmental protection policies are based on the three principles of 'putting protection first and combining prevention with control', 'making the causer of pollution responsible for treating it' and 'intensifying environmental management' (State Council, 1996, p.13). The latter task operates on the system of the 'three simultaneouses', meaning that in any new construction or renovation project, the facilities for preventing and controlling pollution must be designed, built and implemented at the same time as the project itself (NEPA, 1992, p.51). There is a system of pollutant discharge 'permits' system, consisting of four elements: the processing of pollutant discharge statements, setting limits on the relevant total volume of pollution discharge and indicators on overall pollution reduction, issuing verified permits and monitoring their implementation. Local environmental protection bureaux have the power to collect fees and levy fines for excessive discharge of pollutants, the funds so collected being used for pollution control, and in extreme cases to order the closure of a heavily polluting work unit although Jahiel (1997, pp.86-90) makes it clear just how difficult the collection of fees and fines has become. Environmental impact assessments are encouraged for all major construction projects, responsibility systems for environmental protection have been devised and environmental monitoring strengthened (NEPA, 1992, pp. 52-54).

Lotspeich and Chen (1997, p.59) note that while improvements in some areas have taken place, notably in the proportions of waste water treated, industrial dust recovered, industrial solid waste recycled and SO_2 per unit of GDP emitted, the data reveal a 'mixed record of environmental accomplishment' with higher absolute levels of most emissions. NEPA (1992, pp.31-41) identifies several key 'ecological-environmental problems' for the 1990s: the continued overburdening of land resources, grassland degradation, the fragility of forest resources, water shortages, marine pollution and degradation, the loss of species diversity, air pollution caused by coal burning, industrial pollution and urban environmental strain. Clearly, certain issues present themselves as urgent priorities: the need to overcome water scarcity, control over rural industries and slowdown if not halt to the degradation of farmlands, grasslands and forests (Vermeer, 1995, pp.46-48).

5.5 Sustainable Development in China?

Flushed with enthusiasm after the United Nations Conference on Environment and Development in Rio in 1992, the Chinese Government rushed into action and published two key documents in 1994: Agenda 21 (State Council, 1994a) and an Environmental Action Plan for 1991-2000 (NEPA/SPC, 1994). The former, consisting of twenty chapters on 78 programme areas comprises four main sections: overall strategies for sustainable development, sustainable development of society, sustainable development of the economy and resources and environmental protection (State Council, 1994b, p.4). China's first specific and unequivocal commitment to 'sustainable development' is thus made in Agenda 21. As State Councillor Song Jian said to the International Workshop on China's Agenda 21 in 1993 in Beijing,

> We must guide our society and economy along the approach of sustainable development (quoted in State Council, 1994b, p.6).

There is no unambiguous word for 'sustainable' in Mandarin Chinese (putonghua), the most common translation of sustainable development being chixu fazhan. Chixu, however, would more normally be translated as sustained rather than sustainable. The term xietiao fazhan, translates as harmonious development (Edmonds, 1994, p.232). Be that as it may, China's understanding of and approach to sustainable development through the path of rapid economic growth is made clear in Agenda 21:

> Sustainable development is a strategic choice that must be made by both developing and developed countries. For a developing country like China, however, the precondition for sustainable development is development. *The path of relatively rapid economic growth....must be taken.*
> Only when the economic growth rate reaches and is sustained at a certain level can poverty be eradicated, people's lives improved and the necessary forces and conditions for supporting sustainable development be provided. While the economy is undergoing rapid development, it will be necessary to ensure rational utilization of natural resources and protection of the environment (emphasis added) (State Council, 1994a, p.4).

Thus China commits itself to a clear Beckerman-style approach (see Chapter 2) to the environment and makes a specific commitment by the year 2000, not only to have achieved a range of environmental targets [e.g. on industrial waste water, sulphur dioxide emissions, solid wastes, noise levels, forest cover, desertification, water and soil conservation, arable land resources and nature reserves (State Council, 1994a, pp.6-7)] but also to have achieved GNP growth of 8-9 per cent per annum. China's Agenda 21 is a very comprehensive paper of nearly 300 pages explaining how these targets are best achieved.

China's Environmental Action Plan, produced in the same year as Agenda 21 at the behest of the International Development Association (IDA), a branch of the World Bank, details specific sectoral actions plans and prioritises seven key environmental concerns: water pollution, urban air pollution, pollution of industrial toxic and hazardous solid wastes and urban refuse, shortage of surface water, soil erosion, low forest coverage and species depletion (NEPA/SPC, 1994,pp.22-23). Despite the detailed and comprehensive nature of the plan, however, the term 'sustainable development' is fitfully and inconsistently used, unlike in Agenda 21. However, on March 8 1997, Jiang Zemin delivered a speech to the Central Conference on Family Planning and Environmental Protection in Beijing including the argument that,

> overuse of natural resources will make it hard to protect the ecological environment and improve the quality of the environment. If the work on environmental protection is not done well, the environment will be polluted and damaged. This will have a direct impact on peoples' health and living conditions, and will even endanger the survival and development of our descendants (translated from Jiefang Daily, July 12, 1997).

Though the words are not specifically used, Jiang's speech uses concepts wholly consistent with a call for sustainable development as defined in Chapter 2. (I was informed by a Chinese friend who deals in stocks and shares that the price of stocks of companies engaged in anti-pollution technologies shot up on the Shanghai exchange immediately on publication of this speech in July). Clearly, the concept of 'sustainable development' is still important at least for Chinese leaders' propaganda.

5.6 The Future

Smith (1997, pp.3-41) could hardly be more pessimistic concerning the Chinese leadership's responses to the deteriorating environmental situation, particularly given its embrace of rampant consumerism. Certainly there is considerable reason to be alarmed: at all levels of society from TVE managers to national policy makers, the overwhelming imperative is the attainment of narrowly defined economic benefit. Indeed, this is, in some ways, hardly surprising given (i) the currently very low levels of material welfare 'enjoyed' by so many Chinese and (ii) the lack of any easily quantifiable targets at either the micro or macro level in the field of environmental protection for cadres to aim at and win glory by attaining.

Despite the foregoing, however, and despite the doom laden perspectives with which this chapter opened, other scholars are not completely without hope. For Vermeer (1990, p.63) 'one must note that in the 1980s, environmental protection work in China has struck firm roots', for Edmonds (1994, p.257) 'there are some optimistic signs', for Glaeser (1995, p.86) 'China (has) contributed significantly to the eco-development approach' and for Hallding (1996, p.15) 'the most positive and hopeful sign for the future is that the Chinese leadership is increasingly aware of the seriousness of the situation'. Nonetheless it is unlikely that any scholar's optimism is much ahead of Smil who delves into Chinese culture to suggest,

> For any situation there is a saying from the bottomless store of Chinese idioms and proverbs. One hopes that the following would apply in this case: *wu ji bi fan* - when things are at their worst, they begin to mend (1987, p.222).

And despite the successes that he acknowledges China has had in raising environmental awareness and in vigorously executing family planning and anti-poverty programmes, and despite the generally acknowledged seriousness with which the Chinese government has pursued its environmental protection policies and particularly its family planning policies, Vermeer is still able to argue that 'sustainable development is a goal which still does not carry much weight with Chinese economic decision makers (1995, p.49).'

And he goes on to say,

> For a number of political, financial, legal and other reasons, one may expect only minor inhibitions to China's economic growth from environmental and resource constraints. Government and society appear to accept a deterioration of the environment in many areas, particularly rural ones (1995, p.50).

To the particular concerns of Chinese rural areas and to one of the most important policy initiatives to deal with them, CEA, we now turn.

6 Chinese Ecological Agriculture: *shengtai nongye*

6.1 Food Security and the Environment

Over the last fifty years or so, western capitalist economies have effectively dealt with the problem of food security. Since the 1930s with the appearance of large agricultural machinery, the increasing popularity of monoculture, developments in the chemical industry allowing increased use of low private-cost artificial fertilisers and pesticides and other 'scientific' developments in agricultural technologies such as new hybrid cereal strains, there have been rapid increases in farm yields and outputs. Although as late as the 1950s there were fears of food shortages in Europe (one of the initial aims of the Common Agricultural Policy of the EEC was to ensure sufficient food supplies), the most significant political problem associated with the agricultural sector in the '1st World' in the last 30 years has concerned the handling of the huge surpluses produced, so successful, in terms of increased output per unit of land, have improvements in agricultural techniques been. But these techniques, efficient in the use of land, labour and capital machinery, are increasingly polluting, energy-inefficient and wasteful of natural resources.

In the United States of America and Western Europe, there has been in recent years increasing interest shown in the development of more environmentally friendly forms of agricultural production. Unfortunately that growing interest has not been matched, except in the case of a minority on the fringes, by any significant change in the practice of agricultural producers, motivated as they have been by extant government policies such as the high guaranteed intervention prices associated with the Common Agricultural Policy of the European Union to increase farm outputs at minimum short-run private cost to themselves, whatever the long-run consequences to the environment.

In China the human disaster attendant on the 'Great Leap Forward' in the early 1960s continually reminds the current Chinese leadership, if any reminder were needed, of the importance of food security and the last 20 years has seen dramatic increases in total grain yields, the 1997 harvest of 494 million tons (China Statistical Yearbook, 1998, p.386) comparing well with the 1976 harvest of 286 millions (China Statistical Yearbook, 1992, p.328). But as we have seen, output per head has barely kept pace, while the increases have been achieved using 'Green Revolution' techniques, including substantial additional inputs of chemical fertilisers and pesticides and higher levels of irrigation. As a result, the energy intensity of China's agriculture has increased rapidly over the period and the environmental sustainability of recent agricultural development is thus called into question.

6.2 Chinese Ecological Agriculture: an Introduction

According to Luo and Han (1990, p.303) it was Ye Xanji who first proposed the concept of Chinese Ecological Agriculture (CEA) as a development strategy for China's agriculture in 1981. From then on CEA has been promoted by the Chinese government in an attempt to address many key economic, social and environmental issues in the Chinese countryside. In particular, it has addressed three key problems, namely those of: (i) increasing absolute levels of agricultural output to provide security of food supplies (ii) strengthening the economy and increasing rural incomes and standards of living without a crisis of energy generation while at the same time (iii) dealing with the increasingly manifest environmental problems in the Chinese countryside caused by high energy dependency (Zhou Shengkun, associate professor, CIAD, interview, 1995).

Thus Chinese Ecological Agriculture has, since the early 1980s, been one of the primary *systemic* initiatives of China's National Environmental Protection Agency, by which the interrelated problems in the Chinese countryside presented above have been addressed, and remains an important weapon in its armoury of environmental initiatives even today. As recently as the autumn of 1997, Xiong Xuegang, in the officially approved Chinese media, called for China to 'promote ecological agriculture' (Xiong, 1997, p.28).

6.3 The Meaning of Chinese Ecological Agriculture

According to China's leaders, CEA is China's response to the challenge of sustainable agriculture. CEA is, according to Qu Geping 'a mode of sustainable development with Chinese characteristics' *(you Zhongguo de tese chixu fazhan)*. [To label a particular social process of being 'with Chinese characteristics' *(you Zhongguo de tese)* is not an uncommon phenomenon in China. The present reform-and-open economic policy *(gaige kaifeng)* which to all intents and purposes is ushering in healthy dollops of capitalism and free market allocation in China is popularly referred to as 'socialism with Chinese characteristics' *(you Zhongguo de tese shehui zhuyi)*].

As the National Environmental Protection Agency explains in a characteristically *(sic)* elliptical manner, CEA is:

> a comprehensive agricultural production system of intensive management with multiple layers and with multiple structures and functions. It was established on the basis of summing up successful experiences of various agricultural practices applying ecological and eco-economic principles, modern and scientific and technological methods. In a word, eco-farming is a comprehensive agricultural production system which is managed intensively according to the principles of ecology and eco-economics (NEPA,1991, p.2).

It involves an attempt to practise agriculture using many of the techniques of traditional Chinese organic agriculture, in harmony with, rather than in domination of, the natural environment. As such it promotes sound ecological principles, emphasizing traditional practices such as crop rotation, inter-planting and the application of organic manures. As far as possible it emphasizes practices directly of benefit to the environment, including afforestation, the prevention of soil erosion, energy conservation, a reduced dependence on fossil fuels and the utilization of waste.

However CEA goes further than merely applying environmentally friendly principles to extant agricultural practices. Rather, it attempts a comprehensive response to the problems of the rural economy including those of satisfying the increasing material expectations of the burgeoning population and maintaining employment opportunities in the countryside, thereby reducing the already very substantial levels of internal migration from rural to urban areas, indeed maintaining the very fabric of rural communities in the post-Mao Chinese political economy. As a result, CEA puts a great deal of emphasis on the all-round development of the rural economy and specifically on rural industrial employment and income-generating activities. Mao had said 'take grain as the key link, pay attention to animal husbandry, forestry, fish raising and sideline occupations, and develop an all-round rural economy' but, as has been said, the received message appears to have been confined to the first phrase and the rest honoured more in the breech than the observance. For Qu Geping, the message of ecological agriculture was not to be so confined:

> China's eco-farming is a kind of ... 'macro agriculture', based on (the) integrated planning and reasonable arrangement of cultivation, breeding occupations and processing trades. Emphasis is laid on the all-round development of agriculture, forestry, livestock, side-line occupations, fishery and so on, *the integrated operation of agriculture, industry and commerce,* the coordination of the links among various departments of (the) agricultural sector and the practice of (a) diversified economy (emphasis added). (Qu, 1991, p17)

Cheng, Han and Taylor (1992, p. 1135) suggest that the key conceptual underpinnings of Chinese ecological agriculture are fivefold. First, there is a 'holistic' approach to resource use involving attempts to take into account all the existing natural resources in any particular locality and both the production and environmental implications of their use. Secondly, 'stereo' agricultural development is promoted, involving a multi-dimensional use of space and time, an example of the former being the cultivation of rice, azolla (a nitrogen-fixing aquatic plant) and fish in paddy fields, an example of the latter being intercropping. A third component is the promotion of integrated production systems whereby the outputs or by-products of one system are used as inputs of another. Examples of this include the use of organic wastes with biogas digesters in a virtuous circle of material recycling. Fourthly, eco-farming involves environmental management, whereby attempts are made to reduce the use of chemical fertilisers and pesticides, to increase rates of afforestation and to generate power as far as possible, from renewable or continuing resources, for example, from biogas, wind or sun. Fifthly, eco-farming involves diversification of production, combining grain cultivation with animal husbandry, fishing, forestry and other processes with agricultural and non-agricultural rural industries (Han, Cheng and Taylor, 1991, p.1135).

68

The above authors distinguish Chinese ecological agriculture from Western organic agriculture, stressing that in the former, there is a larger, more explicit role for government, a greater prominence of ecological theory, less emphasis on reducing synthetic chemical fertilisers and more intensive enterprise integration and material recycling (1991, p.1139).

6.4 The Legacy of the Past

While CEA has attempted to make best use of scientific advances, it has also been based on sound organic agricultural principles and practices. According to Cheng Xu (1994b, p.415), 'CEA reflects the marriage of oriental wisdom and modern science and technology.'

Systems of organic agriculture are by no means new to China: indeed, as one of the earliest agricultural societies in the world, it has a long history of organic farming. According to China's National Environmental Protection Agency:

> The Chinese nation was the earliest nation in the world to breed and plant glutinous millet, rice, beans and other grain crops. As early as 8,000 years ago, China cultivated millet in the Yellow River valley and planted rice in the Yangtse River valley. Some 4,000 years ago, China was able to plant crops based on soil conditions. As the Zhou Rites noted, the peasants at that time mastered the method of planting different crops based on different soil quality. When weeds grew well in mid-summer, they ploughed fields to turn the weed into manure and tidy up the farmland. This highlighted the importance of planting in the light of local conditions. In building water conservancy projects, controlling floods and improving the soil, irrigation systems were built..... some of them are still operating today. China was the earliest country to breed mulberry silkworm and, during the Han Dynasty, China invented methods of nursing rice for seedlings and growing vegetables in greenhouses. From the Han dynasties (206 BC-220 AD) to the Northern and Southern dynasties (420-589) agricultural cultivation in the Yellow River gradually took the form of a traditional farming system. Centred on ploughing, harrowing and hoeing, farmers commonly applied alot of manure, conserved water and soil, planted green manure crops and used crop rotation to conserve soil fertility. (N.E.P.A., 1991, p.5)

Many of the philosophical traditions in China, including Buddhism, Confucianism and Daoism, despite their differences and conflicts, share a theme stressing the unity of humans and nature. This theme is developed by Yu Muchang, professor of philosophy at Beijing University,

> the idea that man and nature are harmonious has been applied to practical life. For example, it is used in agriculture in China. It is the opposite of the Western idea that man is antagonistic to nature and that man rules nature (Catton, 1992, p.9).

According to Yu, in order to make the most of their land, the Chinese had, by 500BC, developed a system of intensive organic agriculture, based on highly developed systems of recycling, systems which fitted well with the Daoist philosophy which stressed that in order to live at peace with the world it was necessary to work harmoniously with the cycles of nature. According to Catton (1992, p.10),

> Crop and household wastes were composted in the fields, and pigs were used as walking fertiliser factories, converting rubbish into manure. By the Han dynasty (206BC to AD 220), Chinese families were burying their dead with models of pig-pens containing human toilets. These tomb figures were replicas of a system being practised in every farmyard. All household waste was fed to the pigs, which, as well as producing meat, turned the waste into precious fertiliser. The toilet simply allowed the pig to feed on any brown rice too tough for human digestion. The system was a crude but effective way of returning all wastes to the land - with a much reduced risk of transmitting human diseases.

Other techniques of organic farming developed over 2000 years ago which fitted well with Daoist ideals of working in harmony with nature include green manuring, the practice of growing a crop to be dug back in to fertilise the soil, and intercropping. Not only does the latter make best use of scarce land and sunlight, but if nitrogen fixing plants are involved then the soil can be enriched as atmospheric nitrogen is converted into nitrate fertiliser.

The Chinese were also in the forefront of biological pest control. According to Yu Muchang,

> as early as the first century AD, people knew how to use ants to kill pests; in the 3rd century, this method was recorded in books. At that time, people had already learned that ants were able to kill the pests of orange trees. So some made a living by catching ants and selling them in the market to the peasants who grew orange trees (Catton, 1992, p.11).

In addition, other organic systems developed many hundreds of years ago involved the integration of rural industry. An obvious example of this was an ecologically beneficial cycle based around the production of silk (still used in many parts of rural China today). Leaves from mulberry trees would be harvested and fed to silkworms, whose droppings would be fed to fish in fishponds. When the fish were harvested, the ponds would be dredged and the mud on the bottom used to fertilise the mulberry trees. Peasants on the delta of the river Pearl who still practise such a cycle say 'the silkworms are strong, the fish are fat and the mulberry are healthy' (Catton, 1992, p.12).

Many of these organically sound practices stood the test of time. On his trips to the Orient in 1905, F.H. King recorded an impressive variety of organic techniques practised by China's farmers over the centuries. As an example,

> In China, enormous quantities of canal mud are applied to the fields... So, too, where there are no canals, both soil and subsoil are carried into the villages and there they are, at the expense of great labour, composted with organic refuse, dried and pulverized and finally carried back to the fields to be used as home-made fertilisers. Manure of all kinds, human and animal, is religiously saved and applied to the fields in a manner which secures an efficiency far above our own practices (King, 1926, p.22).

6.5 The Legacy of Mao

Despite China's long history of applying agricultural techniques which harmonized human activity with the natural environment, few lessons seem to have been put into practice in the immediate post-liberation China. Indeed, it is clear that Mao, far from emphasising the benefits of man's harmony with nature, stressed the possibility, through appropriate social action, of man dominating nature. Chinese leaders frequently claimed the new socialist nirvana could be created by human action overcoming the forces of nature and as Smil (1987, p.216) suggests,

> the ruling party allied itself very closely with Stalinist orthodoxies, one of them being an attitude of arrogant domination over nature.. Growing up at the time in another Stalinist country ..I vividly recall the repeated propaganda songs about how 'we shall order rain and wind when to fall and when to blow'.

It is therefore unsurprising that some of the greatest engineering projects undertaken in the Chinese countryside in terms of sheer scale are associated with Mao Zedong. Huge projects, many environmentally destructive in the long run, (see Smil, 1987, pp.216-7) were initiated at Mao's behest during the periods of the Great Leap Forward and Cultural Revolution, including forest clearing, irrigation projects and the building of huge hydroelectric dams. The building of the Three Gorges *(san xia)* hydro-electric dam across the Yangtse, the biggest project of its kind in the world, was one of Mao's pet ideas and the decision to go ahead with its construction at the fourteenth Party Congress in 1992, despite enormous worries about its environmental impact (see Edmonds, 1994, pp.83-86, 1996, p.4) owes much to his legacy.

Despite Mao's dictum "take grain as the key link, pay attention to animal husbandry, forestry, fish raising and sideline occupations, and develop an all-round rural economy" (quoted in Hinton, 1990, p.25) statistics suggest that the only message that took hold across post-liberation China was the first, as grain monoculture became increasingly the norm. As has been discussed earlier in this chapter, absolute levels of production were increased in much of the 1950s, 60s and 70s, largely as a result of green revolution techniques which promised man's effective domination of nature in ways which led to the huge importation of energy into the countryside as discussed above and to an increasingly energy dependent agriculture.

The Chinese government became increasingly aware of these problems. According to Qu Geping, in 1991 director of N.E.P.A.,

> China's total output of chemical fertiliser already exceeded 90m tons in 1990; the amount of application ranked first in the world with an average amount of fertiliser application per *mu* ... in China being 13.9 kgs, *which is more than twice the average world level* [emphasis added]. The use of a great amount of chemical fertiliser has brought about a series of problems, for instance, contamination of water bodies, eutrophication of lakes, hardening or crusting of soil, lowering of the quality of farm produce etc. the amount of usage of pesticide is also very big: in some high-yielding areas, the number of applications of pesticide per year is as many as *more than ten times,* with the per mu amount of usage being 1kg. [emphasis added]. The indiscriminate use of pesticide not only contaminates soil, water bodies and a great amount of farm

71

produce... but also results in the loss of ecological equilibrium for farmland (and) the ever-increasing reproduction of insects and pests...thus entering the state of vicious cycles. (Qu, 1991b, p.14)

Qu also recognises other serious problems of the rural environment:

China is also faced with the serious problem of (the) sharp increase of population...(which)...exerts a very great pressure on agricultural production. The per capita area of tilled land is only 1.4 *mu* , ... far below the average world level. However several hundred thousand hectares of fertile farmland are occupied for non-farming purposes every year and about one hundred thousand hectares of fertile land destroyed by natural calamities. The area subject to soil erosion amounts to 1.5m square km. in the entire country, accounting for 15.6 per cent of our territory. One third of the total acreage of farmland ...suffers marked soil erosion; in addition, millions of hectares of farmland and grassland are threatened by desertification and salinization. The forest covers have undergone serious damage (Qu, 1991b, p.14).

And he pulls no punches with regard to the significance of the above:

These problems not only hamper the further development of agriculture and the realization of modernization (but) also in the meanwhile threaten the existence and development of the Chinese nation (Qu, 1991b, p.14).

Professor Cheng Xu of Beijing Agricultural University (in 1996, Head of the Science Department in the Ministry of Agriculture, and in 1997, promoted to the Head of its Education Division) and one of China's foremost propagandists for CEA since the early 1980s summaries China's rural problems by arguing (Cheng, 1994b, pp. 407-15) that Chinese agriculture suffers from a series of 'hard' and 'soft' constraints regarding the use of resources. He suggests there are six 'hard' constraints: (i) the increasing energy intensity of agriculture as a result of the increased inputs of chemical fertiliser, agricultural machinery, fuel and pesticides leading to fertiliser shortage and the need to spend precious foreign exchange on imported chemical fertiliser, (ii) serious shortages of water, drops in groundwater levels and the increasing costs of irrigation, (iii) pollution resulting from the burning of coal to produce chemical fertilisers and from the application of chemical fertilisers and pesticides, (iv) restrictions on the quantity and quality of land resulting from indiscriminate exploitation, overstocking, soil erosion and desertification while the small scale of farming limits its productivity, (v) serious surpluses of labour, leading to low marginal productivity of labour on the land and (vi) severe shortage of investment in agriculture caused, with the exception of the period 1979-85, by the government prioritizing industrial investment and screwing the 'price-scissors' against the farmers.

Among the four 'soft' or 'invisible' constraints, Cheng Xu includes: (i) traditional philosophical constraints, associated with Chinese views on mankind's eventual triumphalism over nature, (ii) institutional constraints, many associated with the reintroduction of family farming, where individual decision making prevails, leading to practices which focus on 'near-sighted profit' (Cheng, 1994b, p.413), the system of promoting cadres on the basis of fulfilment of short-term targets, reinforcing short-termism in resource utilization, certain irrational pricing and accounting policies, such as the externalization of the pollution costs of TVEs and the very low price of water, poor extension services and shortages of infrastructure, (iii) constraints on using sustainable technologies, as a result of farmers having abandoned the 'fine traditions of collecting, composting and transporting and applying manure and dung' (Cheng, 1994b, p.114) and having reverted to 'cheap' chemical fertilisers, and technical constraints on the development of biological pesticides, water saving technology and biogas digestion, and (iv) poverty, leading to indiscriminate felling of trees for firewood, reclamation of marginal lands and overgrazing.

To overcome these constraints, Cheng argues that Chinese agriculture must change its ways and must exclude the model of 'high input-high-output' and replace it with a low input sustainable agriculture (LISA), an agriculture which is best accomplished by the adoption of CEA. Professor Li Zhengfang agrees. While suggesting that Chinese agriculture provides 'cheap' food and raw materials resulting from the 'Green Revolution', he argues this has been achieved at a high price:

> rapid consumption of renewable and non-renewable resources, loss of organic matter and natural fertility, excessive and misuse of water and energy resources, over 600 species of insects and mites resistant to most chemicals, broad-base contamination of the eco-system, the long-term cost of environmental clean-up, displacement of thousands of farm families, loss of biodiversity and local food security.
> None of these costs are now accounted for in the price mechanism... So, in fact the green revolution has created the most expensive food system in the world- if you properly account for resource consumption, environmental degradation and the general health risk.

> So much for the green revolution! (Li, 1994b, p.40)

Professor Li Zhengfang, recently retired ex-Director of the Nanjing Institute of Environmental Science under the direct auspices of NEPA, has, alongside Cheng Xu, been a tireless champion of CEA since its inception in the early 1980s.

6.6 Early Policy Developments

According to Li Zhengfang (1994a, p.281), the development of CEA has taken place in three distinct phases. In the first phase, from 1980 to 1983, there was concentration on academic preparation of basic concepts and major functions and organizing and training scientific and technical staff. In 1980 the government held the first nationwide conference on agro-economics in Yinchuan City, Ningxia Hui Autonomous Region (NEPA, 1991, p.8), in which the term *shengtai nongye* was used for the first time. Two years later an international symposium entitled 'Increasing Agricultural Production Using Ecological Principles 'was held in Kunming and in Guangzhou, jointly sponsored by the Office of the Leading Group of Environmental Protection under the State Council of China and the Environment and Policy Institute of the East and West Centre of the USA which led to the Environmental Protection Bureau (the forerunner of NEPA, now SEPA) beginning to allocate funds to promote experiments in the development of ecological agricultural projects (NEPA,1991,p.8).

It was in the same year (1982) that Professor Bian Yousheng of the Beijing Municipal Research Institute of Environmental Protection set the ball rolling by setting up the first experimental ecological agricultural project at Liu Min Ying, in Daxing County, in the suburbs of Beijing. The construction of such a project in this 'model village' set a precedent for the manner in which ecological agriculture was to be promoted and ushered in the second phase of CEA development, that of experimentation and demonstration, with heavy government involvement in promotion and extension, leading up to the establishment of many demonstration sites and research institutes. In 1984, the State Council pushed the issue further up the agenda by arguing that 'the government departments of environmental protection at various levels should take an active part in popularizing eco-farming' (NEPA, 1991, p.8) and a further conference on agro-ecological environmental protection was held in Wuxuan County, Jiangsu Province in the same year, co-sponsored by the former Ministry of Urban and Rural Construction and Environmental Protection and the former Ministry of Agriculture, Livestock and Fisheries, which made initial arrangements for experimentation and demonstration. 1985 saw the publication by the State Council of Document no.006 entitled 'The opinions concerning the development of eco-farming and the strengthening of work in the protection of the agricultural environment' subsequent to which provincial departments of both environmental protection and agriculture were encouraged to undertake active work in setting up pilot projects. Between 1984 and 1987, NEPA established 19 demonstration projects in 17 provinces, autonomous regions and municipalities.

The third developmental phase of CEA, from 1987 to 1994, involved substantial further demonstration and extension. By 1990, the number of pilot projects had grown considerably, NEPA claiming the existence of over 1300 pilot projects in eco-farming at different administrative levels, 29 at the county level, 138 at the township level and 1200 at the village or farm level (NEPA,1991,p. 9). In 1992, the numbers of demonstration sites claimed had grown to 2,000 (NEPA, 1992, p.23). In 1994, a new project involving seven state ministries and the establishment of fifty demonstration 'ecological counties' was inaugurated.

6.7 Types of Ecological Agriculture

There are, within these numbers, a huge range of types of eco-farming projects, the variety substantially the result of different geographical and climatic conditions and local traditions and practices. NEPA propaganda emphasizes the variety. Some projects have stressed stereo cultivation: Nanhai State Farm in Hainan Island, according to NEPA for example, has been successful in combining the cultivation of forest trees, rubber trees and tea bushes, the former providing a less windy, moist environment for the rubber trees, promoting their growth, while, in turn, the shadow of the rubber trees provides ideal sunlight and a moist growing environment for the tea bushes. And further, the tea bushes protect the soil (NEPA,1991, p34). It is claimed that between 1981 and 1987 Beichuan County, in Sichuan Province successfully experimented with intercropping forest trees and crops (NEPA,1991, p.36), while the Daqiao township in Jiangjin County, Sichuan Province made impressive strides in raising ducks and fish in the rice paddy fields (NEPA, 1991, p.38). And the same propaganda suggests many different systems of material recycling have been adopted: Jixian County, in Tianjin municipality, operates one involving pigs, maggots and chickens (NEPA, 1991, p.46) while Shunde county, in Guangdong province, has another involving silkworms, mulberry bushes, sugar cane and fish (NEPA,1991, p.56) A large number involve biogas as a pivotal element in the recycling.

6.8   Benefits of Ecological Agriculture

Whatever the form of ecological agriculture, its success is judged by the Chinese authorities according to three criteria: environmental benefit, economic benefit and social benefit (Qu, 1991, p.23). This three-fold distinction is not always an easy one to handle- 'environmental' benefits clearly have an economic and/or social component while the fine line between 'economic' and 'social', the former referring primarily to increases in output and income, the latter referring to welfare issues such as housing and education, greater rural employment opportunities and such abstract concepts as a greater sense of socialism and a stronger sense of community, is frequently difficult to draw. Nonetheless, that the success or otherwise of the eco-farming is judged not merely by its impact on the environment but on the economic and social advancement of the people makes clear the importance of CEA in coming up with answers to a multiplicity of problems in the Chinese countryside.

In particular, it is absolutely clear that as opposed to Western 'organic' agriculture, which is adopted for the most part purely on environmental and ethical grounds and which, for its success, is not normally judged on conventional commercial criteria, CEA has to deliver on environmental *and* economic fronts. In a developing nation, this is unsurprising. Where people are scratching to keep body and soul together (and it is important to remember that China experienced generalized famine in the lifetime of anyone aged over 35), issues pertaining to income, output and employment, to consumption and material advancement will inevitably take a higher profile than environmental considerations. Only will the latter be given attention at all if what the World Bank (World Bank, 1992a, p.1) calls a 'win-win situation' presents itself, when particular processes promise both immediate welfare and environmental gains.

6.9   Parallel Developments

While the development of CEA per se continues to take place, notably with the establishment of the fifty 'ecological counties' by NEPA in 1994, a number of related policy initiatives have developed simultaneously. At the end of 1992, the China Green Food Development Centre was established (Li Jianfu, 1994, p.24), and since 1994, the Ministry of Agriculture has been promoting the production of 'Green Food'. Standards of production have been laid down (involving minimal use of chemical inputs), farmers produce crops according to these standards which are monitored and inspected by the Ministry of Agriculture testing agencies at various administrative levels and, if the crops come up to scratch, they are designated 'green', they can be so labelled and sold at premium prices. It is in this regard that market forces are encouraging ecologically friendly production methods.

But also since 1994, organic agriculture, conforming to International Federation of Organic Agriculture (IFOAM) standards, has also been promoted. Professor Li Zhengfang and his team at the Resource Conservation Department at Nanjing Institute of Environmental Sciences, which until then had been responsible for CEA, were transformed to become China's first Organic Food Centre, affiliated to IFOAM in 1994. And in May of the same year, the First Symposium on Organic Farming in China was held at the Centre for Integrated Agricultural Development (CIAD) at Beijing Agricultural University at which Li Zhengfang was a principal mover. Since then, the Organic Food Centre has been responsible for organizing, training, monitoring, inspecting and testing organically grown crops across China, including organic rice, tea, coffee, vegetables, fruits and herbs which, given the seal of approval of 'organic food', can be sold, primarily outside or amongst China's elite or in Western hotels in China, at an even bigger premium than 'green food' of roughly 25 per cent. Organic agricultural standards are tougher than those applied to ecological, or 'green' agriculture, the principal difference being that in the former, absolutely no chemical inputs of any kind are involved. Ecological agriculture involves minimal use of such chemicals, but *some* use all the same, the levels which are deemed acceptable being depending upon local conditions.

6.10 Conclusion

CEA has come to play an important role in the panoply of policies pursued by the Chinese State in dealing with the manifest problems of the rural political economy, including environmental problems, and is increasingly identified, particularly after the 'Earth Summit' in Rio de Janeiro in 1992 as an acceptable model for sustainable agricultural development. Much has been written about CEA in China, often in English translation (Anon, 1990; Li 1991; Tao, 1993; Fang 1995) but frequently in a propagandist vein - often using exactly the same turns of phrase - with little investigation of the specific political-economic conditions of its adoption at local level and the success or otherwise thereof. To such an investigation, we now turn.

7 Research in the Chinese Countryside

7.1 Political Economy

The problematic of political economy predates (neo-classical) economics, which is best understood, not as a branch, or offshoot of political economy, but as a rival problematic. Political economy takes it for granted that economics and politics are necessarily interrelated, that to understand questions concerning the allocation and distribution of material resources, it is necessary to understand questions of *power,* that economic structures and outcomes are based on political relations and vice-versa. Early classical political economists, such as Adam Smith and David Ricardo, while interested in the possibility of maximising material outputs for any given inputs, were aware of the social and political dynamics of their discourse. Both understood capitalism as a social process. It was Smith who formulated 'the iron law of wages' while Ricardo posed the possibility of class antagonisms based on the distribution of value between profits and wages. In the mid-nineteenth century, however, neo-classical economists, while spiriting away any investigation into such philosophical questions as the origin and meaning of 'value' by setting up an enquiry into how prices were determined, also dropped considerations of social formations, classes, i.e. *power,* from its totality of explanatory ideas and concepts as the politically anodine 'laws' of demand and supply took over.

Political economy, while concerned with the implications and outcomes of the processes of production and exchange, thus accepts that *power* threads itself through both those processes: the power of capital, the power of classes, the power of institutions and individuals. Though often associated with Marxism in the twentieth century, political economy is a problematic for all political shades: the approach of Lady Thatcher to the economic problems of England in the 1980s, for example, was as much a political-economic approach as Arthur Scargill's. Will Hutton's (1996) more middle-ground thesis on the state of the British economy in the 1990s is also part of the same political-economic tradition.

7.2 Epistemological and Methodological Issues

The problematic of political economy shares neither the same epistemology, ontology nor methodology as (neo-classical) economics. As Gorelick (1977) argues,

> Methodology.... embodies ways of thinking about reality: these styles of thought are far more subtle than facts but they powerfully shape the selection and omission of facts, the interpretation of facts and the shaping of facts into a theoretical structure: a picture of the world.

Epistemologies assert relationships between subject and object, between the knower and what-there-is-to-be-known. Empiricism as an epistemology asserts the existence of a realm of non-discursive categories independent of and external to the knower, to be grasped, *necessarily capable of being grasped,* by observation. Independent, potentially verifiable 'facts' exist and general truths are merely the representation of these particular facts 'given' to experience. Positivism, as a metatheory of social science, provides an appropriate methodology for an investigation into the social world within an empiricist framework Its principal features, according to many scholars (see Fay, 1975, p.13; Kolakowski, 1993, pp.2-8) are its deductive, nomonological account of explanation and concomitant notion of causality, its belief in a neutral observation language as the proper foundation of knowledge, its value-free ideal of scientific knowledge and its belief in the methodological unity of the social sciences. For positivists, no logical barriers exist between a science of society and a science of nature.

Given the extensive literature and exhaustive debates around positivism which were particularly intense in the 1970s, it is necessary at this stage only to make clear that political economy, and within its tradition this thesis, accepts neither an empiricist epistemology nor a positivist methodology. That is not to deny there is some sort of world, in Bernstein's phrase (1979) 'out there' which needs to be investigated, only to recognise that that world is not static, is created and recreated by the social actors who reside in it, and who make sense of it in different ways. It would also be foolish to disregard all methods of enquiry normally associated with positivism: observation and data collection of various kinds necessarily help to fabricate the story of the process. But it is important to accept that the story *is* a fabrication, hopefully (although inevitably not necessarily) in the more positive meaning of the word. Thus the end result of social science is understanding rather than explanation (in Droysen's methodological dichotomy *verstehen* rather than *erklarung,* see von Wright, 1993, p.11) and 'doing' social science involves a search for signifiers, for signs, which paint in meaning and allow a picture to be built up of the world providing a sensitive interpretation, or Weberian 'empathetic understanding'*(verstehende Soziologie)* of the social processes under investigation (von Wright, 1993, p.24).

In comparison to political economy, economics is conceived by its practioners as a positive science engaged in the business of discovering patterns of determination and regularities in the worlds of production and exchange and neo-classical economics rests its case centrally on the role of the market, with its psychologistic notions of rationality and subjectivist theory of value, as the principal determinant of such regularities. By concentrating on the impersonal role of markets in determining optimal outcomes, the problematic of neo-classical economics frees itself of any uncomfortable baggage associated with questions of conflict and power. Indeed, one of the key ontological principles underlying it is that economics and politics *should* be kept apart, that questions concerning the basis and implications of competitive market equilibria -*positive* questions - can (indeed *must)* be studied in abstract from political structures and processes - which entail *normative* questions. Not to do so would *invalidate* the exercise and would not be economics at all, merely 'old-fashioned sociology' (Blaug, 1970, p.109). In that political economy does grapple with political structures, processes and outcomes, however, its practioners accept that normative questions are at stake, that research interests are value-laden, reject an empiricist epistemology as a result and favour methods of enquiry designed to tease out 'meaning' - qualitative methods - rather than pure reliance on the positivist methods of quantification and number crunching. That is not to say that numbers, where numbers exist, should not be used to build up the picture. Throughout this work continuous reference is made to statistics at all levels of analysis. Without resorting to statistics it is difficult to frame the sort of research questions posed in the earlier chapters, nor analyse them effectively. However, the methodology with which my research questions are handled goes beyond simple reliance on published data and positivist methods of enquiry.

7.3. Political Economy and China

The relevance of political economy in 'doing' research in the PR of China is especially obvious in that there can be no country in the last fifty years where 'economic' processes and outcomes (defined as to do with questions of material production and exchange) are more obviously intertwined with, indeed dependent upon political decisions, upon the very fabric of politics. In no country have economic processes been more overtly 'politicized'. Thus it is fitting to approach research on CEA within a political-economic problematic. After all, this thesis is not designed to understand CEA as a *technical* process, even though the technologies involved are described; rather the technical possibilities associated with them are based on the work of others, of natural scientists' experimentations, and are, to a large extent, taken for granted. What this thesis does investigate, however, are the political-economic, i.e. *social* conditions for their successful adoption and execution in the Chinese countryside.

7.4 Political Economy and Political Ecology

While environmental science is an applied natural science concerned with finding technical solutions to environmental problems (e.g. methods of desulphurization to reduce acid rain) and ecology the science which studies relationships between organisms and their conditions of existence in general terms, human ecology is more specifically concerned with the structure of relations between humans and nature, between society and the natural environment. As Glaeser argues (1995, p.3), to the extent that ecological, i.e. biological and zoological aspects of human ecology are emphasized, then the use of natural scientific methods are appropriate. To the extent that the society-environment relationship is emphasized, particularly when the environmental damage associated with social processes is used as the starting point of research, then

> human ecology becomes *political ecology*, and, as such, its methods
> will be clearly orientated towards social science (Glaeser, 1995, p.3,
> emphasis added).

Following Peet and Watts (1996, p.4), political ecology emerged in the 1970s in response to the theoretical need to integrate land-use practice with local-global political economy. It is closely associated with the work of Blaikie and Brookfield (1987) who suggest that political ecology combines the concerns of ecology with a 'broadly defined political economy' (Blaikie and Brookfield,1987, p.17). Accordingly,

> Environmental problems in the Third World are less a problem of poor
> management, overpopulation or ignorance as of social action and
> political-economic constraints (Peet and Watts, 1996, p.4).

Peet and Watts recognize the importance of Blaikie and Brookfield's pioneering work and in particular of its radical critique of 'the pressure-of-population-on-resources view of the environment' and its affirmation of the centrality of poverty as a major cause of ecological deterioration, pointing to the need for a re-thinking of the concept of development (1996, pp.6-7). Nonetheless, they are critical of its essentially *ad hoc* and voluntarist perspective of degradation, its 'radical pluralism' and its lack of politics and sensitivity to class interest and social struggle (1996,p.8). Rather, Peet and Watts favour a paradigm of political ecology which makes explicit causal connections between the logics of capitalism and specific environmental outcomes, which integrates political action and the institutions of civil society and which questions the radical pluralist approach by investigating:

> why particular knowledges are privileged, how knowledge is
> institutionalized and how the facts are contested (1996, p.11).

This opens a space for what Escobar (1996) calls 'poststructural political ecology' which insists that the constructs of political economy and ecology, as specifically modern forms of knowledge, as well as their objects of study, must be analysed discursively and goes on to argue that:

> It is necessary to reiterate the connections between the making and
> evolution of nature and the making and evolution of the discourses and
> practices through which nature is historically produced and known
> (Escobar, 1996, p.46).

Escobar takes as his initial point of reference the claim made by (Marxist variants of) political economy that capital is going through an 'ecological phase', suggesting that nature is no longer defined and treated as an external, exploitable domain, and that by a new process of capitalization associated with a shift in representation, previously 'uncapitalized' aspects of nature and society are becoming internal to capital. As a primary example, biodiversity is currently regarded as a means whereby genetic engineering and biotechnology can thrive, providing profits for industrial capital. Thus,

> Capital develops a conservationist tendency, significantly different from
> its usual destructive form (Escobar, 1996, p.47).

As a result, the discourse of sustainable development can be embraced, as we have seen in Chapter 2, by hard-headed capitalists as easily as by radical environmentalists. Poststructural political ecology is thus concerned with an analysis of that discourse and of the frameworks within which different 'regimes of truth' contest with each other. Vulgar Marxism is eschewed by poststructural political ecology in favour of Foucaultian discourse analysis, which focuses on the role of language in the construction of social reality, treating language not as a reflection of that reality but constitutive of it (Escobar, 1996, p.46). As Escobar notes,

80

Discourse is the articulation of knowledge and power, of statements and visibilities, of the visible and the expressible. Discourse is the process through which social reality inevitably comes into being (1996, p.46).

Thus, poststructural political ecology nests comfortably within the (post-Marxist) political economic problematic adopted in this work. To the extent that a key tenet of that position holds that social reality is, indeed, constructed through discourse, the methods employed in my research fundamentally reflect this state of affairs.

7.5 Bricolage

Levi-Strauss distinguishes between two different kinds of scientific knowledge, associated with two different levels at which nature is accessible to scientific enquiry,

> one roughly adapted to that of perception and the imagination, the other at a remove from it. It is as if the necessary connections which are the object of all science, neolithic or modern, could be arrived at by two different routes, one very close to and the other more remote from sensitive intuition (1974, p.15).

The 'neolithic', what Levi-Strauss calls 'the science of the concrete' (1974, ch. 1), is thus juxtaposed to modern-day natural sciences. But Levi-Strauss goes on to argue that although the former was inevitably restricted in its ability to achieve results as compared to those destined to be achieved by the latter, 'it was no less scientific and its results no less genuine' (1974, p.16).

Associated with these two kinds of scientific knowledge, argues Levi-Strauss, there are two kinds of searcher after it: the engineer and the 'bricoleur'. While the engineer subordinates tasks to the availability of raw materials and tools conceived and procured for the purpose of the project, the 'bricoleur' operates a large number of diverse tasks with a finite set of tools and materials and 'the rules of his game are always to make do with "whatever is at hand."' (1974, p.17) Indeed,

> The set of the 'bricoleur's' means cannot therefore be defined in terms of a project...It is to defined only by its potential use or, putting this another way and in the language of the 'bricoleur' himself, because the elements are collected or retained on the principle that *'they may always come in handy'* (1974, p.17-18, emphasis added).

Levi-Strauss distinguishes between images and concepts, between the 'signifying' and 'signified' respectively: 'signs resemble images in being concrete entities but they resemble concepts in their powers of reference' (1974, p.18). For Levi-Strauss, the engineer works by means of concepts, the 'bricoleur' by means of signs:

> Thus, both the scientist and the 'bricoleur' might be said to be constantly on the look-out for 'messages'. Those which the 'bricoleur' collects are, however, ones which might have been transmitted in advance... The scientist, on the other hand,... is always on the lookout for *that other*

message which might be wrested from an interlocutor in spite of his reticence. (1974, p.20).

And it is *saturation* into a particular system of signs, of messages, which leads to the successful *socialization* of the bricoleur, to his transmogrification into a scientist.

7.6 Thick Description

The anthropologist, 'does' ethnography using simple methods, including establishing rapport, selecting informants, transcribing texts, mapping, keeping a diary, and so on. (Geertz, 1975, p.6). But according to Geertz, it is not these procedures which define the enterprise, rather it is defined by the kind of intellectual enterprize involved, notably, 'to borrow from Gilbert Ryle, '"*thick description*"' (1975, p.6). Thus while the ethnographer is engaged in a range of routine forms of data collection, he is, in fact, faced with:

> a multiplicity of complex conceptual structures, many of them superimposed upon or knotted into one another, which are at once strange, irregular and inexplicit, and which he must contrive somehow first to grasp and then to render. And this is true at the most down-to-earth, jungle fieldwork levels of his activity...Doing ethnography is like trying to read (in the sense of *"construct a reading of"*) a manuscript - foreign, faded, full of ellipses, incoherencies, suspicious emendments and tendentious commentaries, but written not in conventionalized graphs of sound but in transient examples of shaped behaviour (Geertz, 1975, p.10, emphasis added).

Put another way, thick description involves finding out what you need to know in order to know. The ethnographer is after depth, deep structure, texture. Thus, for Geertz, all ethnographic researches involve *interpretations,* often of a second or third order, and as such, are *fictions,* in the meaning of something made, rather than something false or unfactual. The ethnographer thus acts in the mode of the 'bricoleur', making sense of the world by interpreting signs, signs picked up in case they may 'come in handy' in constructing the fiction. And although as Geertz makes clear, the ethnographic account raises serious problems of verification, or 'appraisal...how you can tell a better account from a worse one' (1975, p.16) he suggests this is really no problem at all, indeed,

> that is precisely the virtue of it. If ethnography is thick description and ethnographers those who are doing the describing, then the determining question for any given example of it... is whether it sorts winks out from twitches and real winks from mimicked ones. It is not against a body of uninterpreted data, radically thinned descriptions that we must measure the cogency of our explications, but against the power of *the scientific imagination* to bring us into touch with the lives of strangers. It is not worth it, as Thoreau said, to go round the world to count the cats in Zanzibar (1975, p.16, emphasis added).

In Henwood and Pidgeon's words, the outcome of qualitative, ethnographic research

82

must be evaluated in terms of its 'persuasiveness and power to inspire an audience' (1993, p.27).

7.7 Research Methods Employed: from Induction to Deduction

Original research for this work took place in the PR China between 1992 and 1997. Though I conducted a large number of interviews with officials and scholars in ministries, bureaux and research establishments in Beijing and Nanjing, the data upon which this thesis is based is primarily the result of fieldwork carried out in seven villages and, to a lesser extent, two counties in different parts of the Chinese countryside. From the outset, I operated within a political-economic problematic, *initially* using methods appropriate to the 'bricoleur' engaging in constructing ethnographic accounts, with the object of producing thick (rather than thin) description.

At the outset, it must be acknowledged that the actual modus operandi of my research involved making a number of *short* visits to several villages over a period of time (for example, I made nine daytime visits to Liu Min Ying over six years, while I visited Xiao Zhang Zhuang for stays of three-to-four days over five consecutive summers). There are clearly advantages of making short visits: in particular, they allow valuable comparison both across China and over time to be made. To the extent that Clifford Geertz's anthropological methods assume *immersion* into a community over a long period of time, however, it is inevitable that my research methods failed to do justice to Geertz's ideal. Though my goals *were* to produce thick description, I am aware that the description I have ended up with is at times thinner than either Geertz or I would have liked.

According to NEPA there were more than 2,000 CEA demonstration sites by the early 90s (NEPA, 1992, p.23), some at the farm level, some at the county level, most at the village level. My researches at NEPA, the Ministry of Agriculture and at the key.institutes in Beijing and Nanjing suggested that those sites represented a large variety of forms of CEA adopted and an equal variety in the success of their adoption. NEPA's own published work on CEA presents a bewildering variety of eco-farming experience (NEPA, 1992,a, pp.31-86). It is thus difficult to conceptualise a 'typical' eco-village.

The villages visited in my study were not chosen randomly: each was a model village (see 7.10 below) which had already to a greater or lesser extent successively adopted CEA. Four of the seven villages had won the UNEP 'Global 500' award, two further villages were nominated in 1995 by their provinces for that award and the other was a nationally well-known model village for a range of reasons, not least ecological agriculture. What is certainly the case is that these seven villages were on the 'successful' end of the spectrum.

Though working within Chinese model villages has its problems (see 7.10 below) the advantage of doing so in this regard is two-fold. First, it is *easier* to research good things rather than bad (true everywhere, but particularly in China). Secondly, to the extent that the claims of the village to forms of environmentally friendly economic development had in every case been checked out beforehand at least by provincial experts, and in the case of four villages, by national and international experts, I was able to take the *technical* issues to some extent for granted and to use my time mainly in teasing out the *social,* political-economic, dimensions of the processes involved. My raw materials were a note book and an interpreter and my methods were typically those of the ethnographer, initially building up good relations, sorting out interviewees, carrying out semi-structured interviews, walking around, observing, mapping historically, writing up my journal. I did not go around with a meter testing soil quality.

The amount of time spent in each village varied: in one, near Beijing (Liu Min Ying)I made nine day-time visits between 1992 and 1997, in another (Xiao Zhang Zhuang, I stayed for periods of three days on five occasions in each of the summers from 1993 to 1997. I visited and stayed in He Heng four times in the summers of 1993, 1995, 1996 and 1997 and visited Dou Dian three times, in the summers of 1995, 1996 and 1997. In each of two other villages, however, Tou Teng and Qian Wei, I visited only twice, in the summers of 1995 and 1997 and I stayed for four nights in Tie Xi on one occasion only during the summer of 1995. In all but one (He Heng, where we stayed in the county hotel, a few kilometres from the village), my interpreter and I stayed in the village 'guesthouse' Thus the thickness of the description varies with the village, with the amount of time spent in it and the number of visits made over time.

And the thickness of the description varies not only as a result of the above differences but as a result of the different opportunities made available to me in them. In every village I was able to talk to the senior leaders including the Party Secretary *(shuji)* but opportunities to speak to ordinary villagers varied considerably. Each interview was structured to the extent that I asked similar questions of each interviewee of similar standing, although the outcomes often differed as I preferred to follow up interesting responses rather than religiously get to the end of my list of questions. In every case, I attempted to *maximize* the number of message-giving signs with the object of obtaining at least *some* thick rather than alot of thin description. I preferred long interviews with a few people rather than short interviews with many. Particularly given the difficulties of research in China (see 7.10 below), short interviews would rarely be of great use anyway - responses would be stock responses - and the most productive would be those of long duration -sometimes of three to four hours - where the interviewee would find him or herself so involved in the story that time flew by for all concerned.

Initially, I began as a bricoleur: I did not know what I needed to know in order to know. When I first visited Liu Min Ying in 1992, I knew little of the nature of Chinese villages (my desktop research had not then stretched either to Myrdal's, Oi's or Potter and Potter's seminal works [1965, 1989 and 1990, respectively], a deficiency later remedied), I had no mental picture of how they would be, no sensitivity to the power relations within them, no idea that patterns of land ownership and usage differed from one village to another, only a sketchy idea of the implications of modern Chinese history on village life, certainly no idea that some villages still operated as cooperatives, no previous direct experience of agricultural activities and only the faintest notion of what biogas was and how it could be generated. In the early stages of my research therefore, I, as bricoleur, conducted research *inductively,* trying to read signs and interpret meaning with the object of building up hypotheses, theories, *concepts,* as Levi-Strauss denotes them.

But as the research went on, so I moved imperceptibly from bricoleur to engineer, from the

84

former, operating inductively, to the latter, acting *deductively*. Instead of looking for signs through which the fiction could be constructed I began to develop hypotheses, concepts, which related one sign to another and through which I made sense of the social processes involved. 'Reflexivity' began to shape and constitute the object of enquiry (Henwood and Pidgeon, 1993, p.24). Towards the end, I found myself using my research time to *test* those hypotheses against empirical evidence, judging and interpreting social processes primarily in their light. To give a specific example, one of the most important concepts built up by the initially random data that I, as bricoleur, observed, was the concept of *leadership* and its importance in determining outcomes. (An understanding of leadership in Chinese villages needs particularly thick description, indeed far thicker description than my short visits, however frequent and intense, allowed). By the end of the research I was essentially testing my hypothesis of *the importance of leadership* wherever I could.

The method I have just described follows closely the method first conceptualized by Znaniecki in 1934 (Bulmer 1984, p.249), that of *analytic induction,* a procedure intended to maintain faithfulness to empirical data while abstracting and generalizing from a relatively small number of cases.

> Its aim is to 'preserve plasticity' by avoiding prior categorization (Bulmer, 1984, p.250).

Analytic induction is contrasted with *enumerative induction,* which involves the collection of a large number of cases and once they are assembled, the formulation of categories and empirical generalizations on the basis of the data available. While the latter form of induction abstracts by generalizing, *analytic induction* generalizes by abstracting (Bulmer, 1984,p.250). It abstracts from a given set of data the features that are essential, and generalizes them. Thus Znaniecki advocated a process which involved first discovering which features in a given set of data are more or less essential, then abstracting these features with the assumption that the more essential are those that are more general than the less essential and must be found more frequently, then testing this hypothesis by investigating classes in which both the former and latter are found and finally organizing these classes into a 'scientific' system based on the importance of the respective features in determining them. However Robinson (1969, p.204) explains,

> the method leads in practice to the isolation of the necessary and sufficient conditions for the phenomenon. It leads to the isolation of characters of such a nature that when the characters are all present, the phenomenon occurs and when the characters are not all present the phenomenon does not occur. However the method does not lead us to the conclusion that these characters are 'essential' apart from the fact that they are necessary and sufficient as operationally defined.

Accepting this caveat, research for this thesis and its conclusions have largely followed Znaniecki's advocacy.

7.8 Grounded Theory

To the extent that theory has emerged from my research in the manner described above (and despite some differences between the following and analytical induction, see Bulmer, 1984, p.255), what has emerged is *grounded theory.* In coining this term, Glaser and Strauss (1967) were attempting to free social researchers from the straitjacket of a few 'grand' theories. In their work, grounded theory refers to theory generated in the close inspection and analysis of qualitative data, produced as a result of research undertaken without any strong *a priori* theory. Thus, in the attempt to make sense of a large amount of unstructured data, in Levi-Strauss's language, as the bricoleur constructs new concepts from signs, the researcher 'is endowed with maximum flexibility in generating new categories from the data' (Henwood & Pidgeon, 1993, p.21). Following Henwood and Pidgeon,

> this is a creative process which taxes fully the interpretative powers of the researcher, who, in Glaser and Strauss's terms, *fit* the data well. Success in generating grounded theory which is faithful to the data depends upon maintaining a balance between full use of the researcher's own intellect and this requirement of fit (1993, p.21).

This crude description of the generation of grounded theory is clearly problematic in that researchers do not come to the field as 'tabula rasa' and to some extent they define what counts as legitimate data by reference to theory. As Henwood and Pidgeon suggests, this raises the question, what grounds grounded theory? (1993, p.22) As with the method of *analytic induction,* there is always some interplay between data and theory, between signs and concepts, there is always, according to Bulmer (1979, see Henwood and Pidgeon, 1993, p.22), a 'flip-flop' between ideas and research experience. However to the extent that the research for this thesis began without any overt *a priori* theorizing and continued through the inevitable flip-flop of signs and concepts to produce sufficiently thick description to allow theory to be developed which can persuade an audience, I have followed the precepts of analytic induction and hope, in Glaser and Strauss's words, to have produced theory which 'works' (Henwood and Pidgeon, 1993, p.24).

7.9 The Rural Development Tourist

Chambers (1983) soberly brings Third World researchers up short by reminding them of the dangers of rural development tourism, the phenomenon of the brief rural visit, its problems 'often 'too near to the end of the nose to be in focus'' (1983, p.10). Rural development tourists, despite widely varying characteristics in other respects, usually have, according to Chambers, three things in common, and were certainly true about me:

> they come from urban areas; they want to find something out and they are short of time (1983,p.11).

86

Apart from any other problems associated with it, rural development tourism frequently involves six forms of bias: (i) a spatial bias (researchers stick near tarmac, urban areas and roads), (ii) a project bias (researchers go where particular projects funded by officials are taking place), (iii) a person bias (researchers speak overwhelmingly with 'elites', men, users or adopters, with those who are active, present and living), (iv) a dry season bias (researchers avoid the rainy season in which hunger, sickness and other difficulties are more prevalent), (v) a diplomatic bias (researchers want to be polite) and (vi) a professional bias (researchers' professional training and values may make it difficult for them to gauge the complete picture) (Chambers, 1984, pp.13-23). My experiences suggest that it is difficult not to be guilty of at least some if not all of these biases. However, in defence of my research, I would argue the following:

Regarding (i), although I visited the village in my study nearest Beijing (Liu Min Ying) more often than any other, the villages I stayed in longest were Xiao Zhang Zhuang - miles from anywhere in the heart of rural northern Anhui which I arrived at, after journeys from Beijing of twenty-four hours or more by train and uncomfortable rural bus - and He Heng, so cut off in central Jiangsu province that on one occasion I was only able to reach the village by river. Indeed, so difficult was it to get even to Liu Min Ying on the first occasion, though only 40 kilometres from Beijing, that it took five hours and I ended up arriving on the back of a horse-and-cart. When I arrived at Xiao Zhang Zhuang for the first time with my interpreter in a dilapidated, uncomfortable bus from the nearest town seven hours away, tired, dirty and dispirited, I realized my intermediaries had clearly oversold me to the extent that we were feted like the king and queen of England, welcomed by provincial as well as county and township officials, sung to by the little schoolchildren and provided with delicious food. As I gave a speech and signed my name in the visitors' book immediately under Mostafa Tomba (the then director of the UNEP, who had visited two years previously), I mused that M. Tomba and his team from the United Nations had probably not arrived on a bus full of chickens.

Thus, while there are no doubt still more rural locations in China, more cut-off from urban centres that I could have visited, I believe my research was not guilty of a serious spatial bias and that the villages I visited represented, spatially, a reasonable cross-section. As it was, the discomfort involved in reaching the villages alarmed my urban-dwelling interpreter to the point of her frequently threatening to go on strike. Meanwhile, living conditions in them, particularly in the height of summer, in bedrooms with no air conditioning, poor plumbing and incessant mosquitos and other insects and animals to contend with, were not normally of the type associated with soft urban living.

Regarding (ii), my research did inevitably suffer from project bias, but to the extent that I was not attempting a thoroughgoing ethnography of the villages concerned, but an understanding through ethnographic methods of the social basis of the adoption of CEA, project bias was *inevitable* and not damaging.

Regarding (iii), given the *impossibility* of visiting any village, or indeed any work-unit in China, without some form of official approval, it was inevitable that I should have contact with village cadres. However, I felt it was *important* to do so. Indeed, to the extent that the village leaders, particularly Party secretaries, were empowered to make key decisions, and were often prime movers, I felt it important to speak to them at length. In all cases they were older rather than younger and were in a position to give historical overviews. They were, with one exception (the leader - *cunzhang* - of Qian Wei), men, although I did meet female cadres in three other villages. When given the opportunity to speak to villagers, I always asked (although I was not always successful) to speak to women. But in civil society in China, almost all decision-makers are men.

Regarding (iv), in that I did not visit any sub-tropical areas of China, I do not believe my summertime bias was of any particular significance. Indeed, the almost unbearable heat in parts of South and Central China during the summer did not put the villages in any especially favourable light.

Regarding (v) it is exceedingly difficult to go anywhere or do anything in China without having or maintaining good relations *(guanxi)*. Also given the importance of 'face' in China, to enquire too deeply into failure or make criticisms would be socially unacceptable and invite an adverse reaction - including likely withdrawal of *guanxi*. The problems associated with this bias were to some extent mitigated by the fact that I was largely investigating 'success', rather than failure. To the extent that politeness was necessary to elicit any thick description at all, however, I was as polite as I could be in all circumstances, even when this involved drinking large quantities of liquor at the end of a meal.

Regarding (vi), as 'bricoleur' initially, I had no particular professional bias - indeed, I was not there as a specialist although I was often asked to give my advice as though I were. I always declined on the grounds that I was there to find things out from them, not to tell people what to do.

7.10 Particular Problems of 'doing' Research in the Chinese Countryside

Wolf (1985) entitles her chapter on the doing of research in China *Speaking Bitterness*. The interested observer, therefore, need read no further than the title to get a flavour of what such an activity can be like.

China is a country in which it is difficult for anyone to do research, but particularly for foreigners. At the most basic level, there is the problem of language. After all, the Chinese themselves frequently find it difficult to talk to each other - the residents of Beijing speak a different dialect from the residents of Shanghai and a completely different oral language from those of Guangzhou. But, additionally, the older farmers of northern Jiangsu speak a different dialect from those of southern Jiangsu and neither necessarily speaks Mandarin *(putonghua)*. To the extent that this is mirrored even in just about every Han populated province in the country, not to mention those communities where Chinese -*Hanyu*- is not spoken at all, it is clear that a researcher may well have considerable problems, even one competent in Mandarin, at establishing dialogue in more than one place.

On all my researches into the Chinese countryside, I was accompanied by Mandarin-speaking interpreters. To the extent that all - except one - was extremely competent, I have little doubt that my interviews, though mediated through another person, mostly the same person, were faithful to my objectives. On more than one occasion, however, there was a need for a second interpreter to translate Mandarin into the local dialect and vice-versa. Working through an interpreter has its advantages. Since I was not always aware of the cultural rituals necessary for discourse with Chinese, interpreters were often able to rephrase too pointed a question into a form polite enough to ask or even embellish my words with suitably empathetic body language. The interpretation, therefore, was not purely linguistic. On a more technical point, working through interpreters gives everyone involved (except the interpreters) more time for thought and reflection. But to the extent that doing so necessarily involves researching at a distance and demands not only one's own dedication and enthusiasm but another's, means that researching can be a hit-and-miss affair. On only one occasion did I miss significantly. On that exceptional occasion, I ended up with an interpreter for five days who knew nothing and cared less about ecological agriculture, did not bother to learn the - many - technical terms involved, got drunk with the village leader in an important *guanxi* - establishing banquet and urinated in front of the official village car before collapsing into unconsciousness.

In common with many developing *and* developed countries, official statistics cannot always be relied upon in China. That is not to deny the existence of a substantial bureaucracy in China (viz. the State Statistical Bureau founded in 1952) devoted to the production of comprehensive statistical material. Indeed, as Clarke suggests,

> China has had a well-established and comparatively efficient land-recording and administrative machinery for a long period. Overall it shares a history of relatively accurate rural record-keeping in common with other large-scale bureaucratic Asian states such as Pakistan and India..... Large-scale census and survey work is now carried on routinely by the administration in China (1992, p.217).

Nonetheless, Clarke's discussion of the problems of research design in rural involves reminds us of the inevitable institutional factors impinging on the accuracy of survey data. Indeed, he suggests that

> the range of variation in the types of survey error suggest that few quantifiable correctives can make the data uniformly accurate (1992, p.217).

And few scholars would dispute, in a country so *politically charged* as China has been since Liberation, that official statistics have been frequently subject to political manipulation. There can be little doubt that the magnitude of the disasters and chaos experienced in the periods of the Great Leap Forward and the Cultural Revolution were associated with inaccurate data reporting and handling. As Clarke writes,

> In 1957, the (State Statistical) Bureau became subject to political criticism and during the Great Leap Forward became subordinate to populist and ideological reporting requirements. The Marxist notion, contra Malthus, that production increases would continue to outstrip population increases, acted as a pressure on statistical bureaux: this resulted in distortions in publicised data. There were some reforms in

1961 which placed restrictions on the Party or administrative departments altering figures; but during the Cultural Revolution the earlier problem of political distortion reappeared and there were even absences of data for a number of years altogether. The State Statistical Bureau did not begin to recover until 1978 but by 1982 had still not reached its earlier staffing levels (Clarke, 1992, p.217).

And since 1982, there has been no great cause for comfort with regard to the accuracy of data. Indeed, there are so many reasons why under-reporting or over-reporting is commonplace that statistical data must be treated with extreme caution. A few examples will suffice to illustrate the general problem. As Muldavin explains, the size of the 'bumper harvest' of 1984 can be at least partly explained by the historically high prices available to peasants:

> Peasants sold surpluses normally held in reserve. And the statistics for the grain harvest were based on deliveries of grain to grain stations, not on field site testing of production levels. (1997, private conversation).

Other examples Muldavin cites for inaccurate reporting include the simple desire to please on the one hand and to make claims upon state resources for relief on the other. As illustration of the latter, Muldavin remarks,

> Peasants develop indirect forms of resistance. In a state which declares that each person contributes according to his/her ability, disasters are legitimate calls on the state's forbearance with regard to taxation - the state has little moral choice but to come up to the aid of the stricken communities. Despite sustained favourable weather during the mid-1980s there was a 'strange' increase in peasant declarations of natural disasters as village cadres mastered the art of 'poor-mouthing' to strengthen their claims. Thus between 1981 and 1989 natural disaster claims were made for seven of nine years on behalf of villagers in Songhuajiang township to justify lower annual quota requirements, disaster relief, development aid, cheaper and more abundant inputs and cheap or free credit (1996a, p.250).

As explained in Chapter 4, Lu Liangshu (1995, p.1) of the Chinese Ministry of Education dampens down concern over problems for China of being able to feed itself by admittting the gross under-reporting of the real amount of farmland and per area yield.

Thus there are good reasons to be sceptical of official statistics, however much they form the backdrop of research activities. Chambers' words are apt in this regard:

> Findings printed out by a computer have a comforting authority. The machine launders out the pollutions of the field and delivers a clean product, which looks even cleaner and more comfortingly accurate when transferred to tables and text. These 'findings' are artifacts, a partial, cloudy and distorted view of the real rural world (1983, p.54).

Most of the statistical data reported from the villages in my research are word of mouth from the village leaders. To the extent that these statistics no doubt conform to the statistics put in documentary form by the same leaders, there is no problem of verification. But to the extent that the figures may well be fabrications (i.e. *false*) in the first place, there is.

China is a totalitarian state where discourse runs from top-down, a country where people - at least people without a great deal of power - are unused to asking (or answering) questions. It is moreover, a country where, despite nearly twenty years of reform-and-opening to the outside world, foreigners are treated simultaneously with a mixture of respect, friendliness, suspicion and resentment. Thus for a village cadre or farmer to be confronted with questions by a foreigner is itself a very strange state-of-affairs. In a totalitarian society where there is a politically acceptable, official line on just about everything, it is likely that answering them will provoke a degree of anxiety unimaginable to those living in more pluralist regimes. This was the case even when I was not accompanied by a 'minder', but still more so when I was. In such circumstances, the establishment of relationships, of good *guanxi*, is absolutely essential if the researcher is to elicit anything beyond the stock response. I was often extremely fortunate in finding myself in villages as a result of contacts who themselves had good *guanxi* and unimpeachable integrity. But even then, work had to be done to establish myself and I was most successful in doing so when I was able talk over a long period of time and/or visit on a number of occasions. Trust built up with some degree of friendship over time. And as that trust developed, the story often changed. One of my most important sources in Beijing told me on our first meeting that the land reforms had been good for Chinese agriculture, two years and four meetings later he confided they had been disastrous both for the long-run health of agriculture and of the rural environment.

As Potter and Potter (1990) suggest, 'reality is explained through a shared public morality' and it is clear that a prerequisite for arriving at the 'truth' is a degree of sensitivity, even empathy, with that morality. Without it, the opportunity of going beyond the 'line' is closed off and description risks not only being thin but false. And the danger of being fobbed off with the official line is that much greater in a Chinese 'model village'.

Officially approved policy and practice in PR of China is maintained and reinforced by the concept of the 'model' *(mofande)*, be it a model bus conductress, model worker or model village. The apotheosis of the latter was the campaign to 'learn from Dazhai' (see Chapter 3). That the Chinese propaganda machine was efficient in promoting Dazhai has never been in doubt, but the degree to which Dazhai's achievements owed more to propaganda than reality certainly has (see Hinton, 1989, pp.124-139). As a result of that legacy and natural suspicions faced with an *advertizement* of what is supposed to be good, foreigner researchers normally treat the claims of model villages with great care.

The research for this work was based on the experiences of such villages; in the context of the question set up, however, this was inevitable, in that any village that successfully set up CEA *in so doing,* became one. Liu Min Ying, in Daxing County, Beijing Municipality, for example, is probably the best known eco-village in China, it was the subject of one of the very first ecological agricultural construction projects and was the first Chinese award winner of the UNEP 'Global 500' citation in 1987. It has thus attracted many famous visitors from home and abroad and was a common destination for delegates to the UN Womens' Conference in Beijing in 1995. But its model status was obtained neither by accident, divine intervention nor osmosis, but rather by taking an imaginative decision involving high risk and initially a degree of political obloquy and by hard work. The same is true for every village visited: model status was not conferred on the village before it had become an exemplar, but *afterwards*. That is not to say that its claims are necessarily true nor that the need to justify continued model status may not require manipulation and/or economy of the truth. But in every case, the village was a functioning 'real' village both before and after the adoption of CEA (with, as we shall see, varying degrees of success). None was a Potemkin village. Of course, I am aware that the Chinese do not only necessarily want to fool foreigners especially, that where there is a charade, the Chinese are 'far more interested in fooling each other and collectively maintaining the charade' (Zweig, 1989, p.14). That any charade is not involved I cannot be *sure* , but, if it has it will have been played convincingly and over a length of time and has fooled not only me and my interpreters but township, county, provincial, national and (in four cases) international arbiters.

The phrase 'model village' may also mislead the imagination of those ill-acquainted with China. Liu Min Ying is perhaps the most famous eco-village in China, yet it remains almost impossible to reach without prior knowledge of its *precise* whereabouts. While my first visit there (in 1992) took a five hour journey from Beijing, involving four buses, a tram and a horse and cart, my last journey (in 1997) by taxi was almost as wearing in that the taxi driver could not find anyone to give helpful directions to the village even when we got to within a few kilometres of it. There are no signs, no obvious 'signs' of it being a model village until the village itself is reached and even then very few. The sign 'welcome to Liu Min Ying', written in English was erected only in 1995 and in 1996 was already in a dilapidated state. In none of the other villages was the situation different. Indeed to the extent that these *were* exemplars, the question 'what on earth could the others be like?' might well occur to interested observers.

A problem of any research into exemplars is the lack of 'controls'. When I voiced my concern about the problem of researching only apparently environmentally successful villages in China within any controls, it was confidently suggested to me by an experienced China scholar that, as far as China and the environment was concerned, this wasn't a problem. 'Just use the rest of China as controls' As mentioned earlier, it is almost impossible to research *failure* in China in any case: on one occasion I made a successful visit to Liu Min Ying with a producer and camera team from a Chinese television network - they were doing a series on environmental issues for International Environment Week in 1995. The following day, the same team went to a well-known Beijing Steel Works, not a place renowned for environmental sensitivity, and after asking a few somewhat searching questions concerning environmental protection, the television producer was accosted by guards and held for two hours before his bosses were able to intervene. His film was confiscated.

Of course, as with all fieldwork in the Third World, there are enormous logistical problems of doing research in China. It is a very large country, with enormous extremes of temperature, with an unreliable telephone system and a major problem of transportation, particularly to places off the beaten track. Researching in China is made that much more difficult by the *absolute* necessity of making good contacts, having good *guanxi*, and, as a result, having influential support for one's activities. Politically authoritarian states are not famed for taking a lax view of foreigners snooping about and China is no exception.

7.11 Conclusion

In the words of Jan Myrdal when introducing his 'Report from a Chinese Village' (1965, p.xiv), the usefulness of my research and the validity of my ideas must ultimately 'rest on my honesty' and will not 'try to hide the fact behind an objectification which is misleading.' And given the difficulties of doing research in China explained above, the much more recent words of Gordon White (1993) are particularly apt. For apart from having enormous energy, dogged resilience, a thick skin and good fortune,

> the scholar must borrow the skills of the detective and investigative journalist, drawing together a wide variety of sources and reading as often *between the lines* as along the lines. (Research in China) is an exercise in analytical demystification (1993, p.13, emphasis added).

8 Case Studies

8.1 Introduction

In order to observe the performance of CEA in villages and counties in which it has been adopted I travelled to seven villages and two counties across the Chinese countryside. I visited four winners of the UNEP Global 500 Award: *Liu Min Ying* in Daxing County, Beijing Municipality on nine occasions between 1992 and 1997; *Xiao Zhang Zhuang,* in Yingshang County, Anhui Province each summer between 1993 and 1997; *He Heng* in Tai County, Jiangsu Province in the summers of 1993,1995, 1996 and 1997; and *Teng Tou* in Fenghua county, Zhejiang Province in the summers of 1995 and 1997. I also visited three further villages that had adopted CEA: *Dou Dian* in Fangshan district, Beijing Municipality in the summers of 1995, 1996 and 1997; *Qian Wei* in Chongming Dao, Shanghai Municipality in the summers of 1995 and 1997; and *Tie Xi* in Mishan County, Heilongjiang Province in summer 1995. Additionally, I visited two ecological agricultural counties, *Da Zu* in Sichuan Province and *Simao* in Yunnan Province in the summer of 1995. My researches therein are described in the case-studies following.The locations of these villages and counties are illustrated below:

Liu Min Ying, in Daxing County, Beijing
D: Dou Dian, in Fangshan District, Beijing
X: Xiao Zhang Zhuang, Anhui Province

T: Teng Tou, Zhejiang Province
H: He Heng, Jiangsu Province
Q: Qian Wei, Chongming Dao, Shanghai
TX: Tie Xi, Heilongjiang Province
DZ: Da Zu, Sichuan Province
S: Simao, Yunnan Province

8(1) Liu Min Ying

8(1) 1. Introduction

Liu Min Ying is a small village of some 898 people in 240 households in the suburbs of Beijing, in Daxing County. Standing on the North China Plain, it lies 30 miles to the south-east of China's capital close to the recently opened highway linking Beijing with Tianjin. Despite its relative proximity to two centres of large urban populations, Liu Min Ying has an entirely rural location, surrounded by other villages similar in size. Access is not easy except by car, although with the rapidly increasing prosperity of the region, many more cars make journeys to Liu Min Ying now than a few years ago and travelling to Beijing for villagers is becoming more routine than it was, even very recently. But drivers must know where they are going - road signs to the village are at a premium and few people can give accurate directions to the village even within a few kilometres of it. Buses from Huancun, the county town of Daxing, are infrequent making life difficult not merely for the carless visitor to get to the village but also for the local people, young as well as old, to venture too far from their homes. The village owns 80-odd motorised vehicles (including tractors, lorries and taxis) which are used intensively in various forms of local economic activity; for most people, however, bicycles and bicycle carts remain the most popular, indeed the only methods of getting around. In that regard, Liu Min Ying is a very typical Chinese village. It remains untypical in many other respects.

8(1) 2. Recent History

The origins of Liu Min Ying, according to stories handed down by word of mouth, go back to the late Ming dynasty, that is, to the late 16th early 17th centuries, when the inhabitants of 72 villages in Shaanxi province were forced by famine to migrate from their homes and to travel by foot and without livestock or possessions the several hundred miles eastward to the current location of Liu Min Ying. This site was attractive because of the proximity of the river Fen and the availability of sufficient underground water for successful grain cultivation.

At liberation in 1949, Liu Min Ying was composed of approximately 100 households from relatively few extended families. The Zhang family was the largest; even today Zhang is the most common family name. Before liberation, there were two landlord families and three rich peasant families whose landholdings were large enough to necessitate hiring the labour-power of poor peasants who owned no property. The middle peasants relied largely on their own resources. During the first stage of land reform immediately following liberation, the land of the landowners and rich peasants was divided, if not entirely equally, at least more equitably, amongst the poor peasants and the landowners were forced to accept reeducation and reduced to working the fields alongside other village members. There appears to have been no great hostility to the landowners. Indeed there was some sympathy for them and they were not personally attacked either at liberation, or later. The economy of Liu Min Ying during those first few years after liberation until 1953 when the initial land reform came to an end, was one based almost entirely on private ownership of land, although a degree of mutual aid developed, including the sharing of draft animals.

In 1953 the second stage of land reform was ushered in with the arrival of the cooperative movement. The peasants' households, their lands and assets, were organised into cooperatives (*jiti*): each primary cooperative in Liu Min Ying involved about 30 people in seven or eight households, in which land was collectively owned and output was distributed according to work performed. It was at this time that the system of work-points emerged. By the time of the Socialist High Tide, in 1956, the village unit itself disappeared, all primary cooperatives being grouped together in a higher level cooperative (*gaojishe*) incorporating many of the erstwhile neighbouring villages.

In 1958, Mao's Great Leap Forward was promulgated. Neighbouring cooperatives were merged into brigades (*dadui*), which were, in turn, lumped into much larger peoples' communes (*gongshe*), made up of several thousand households, which collectively owned the land and all other means of production. Liu Min Ying became one of the 25 brigades making up the peoples' commune of Zhang Zhi Ying. It was at this time that the greatest degree of social levelling was attempted: families were discouraged from cooking at home, for example, and forced to eat in the communal canteen which lasted from 1959 to 1961. But other aspects of the Great Leap Forward, such as the production of pig-iron in backyard furnaces, did not take place in Liu Min Ying. This might help to explain why there was no *pronounced* economic stress there in the early years of the 1960's immediately after the Great Leap, as existed in other parts of the Chinese countryside, despite three years of natural disasters. Though there was hunger, there was no generalised famine. Nevertheless, conditions were hard, peasants worked on the still unflattened land with little or no interest, and grain output was as little as 100 kgs. per *mu*. Meanwhile, since all forms of household level economic activity were outlawed as `capitalist tails' (*ziben zhuyi weiba*), standards of living were accordingly very low. There was insufficient food and what there was displayed no variety. Peasants ate steamed cornflower bread (*bang zi mian wo wo tou*), a food now used for animal fodder. There were clothes shortages, no electrical appliances and the traditional *kang* provided the only heating in the mud-brick homes.

But the early 1960's was a formative period for Liu Min Ying as the leadership began to address the food problem and to consider ways to improve the brigade economy. It was decided to raise both land and labour productivity. As a result, land was flattened, labour divided into groups and work evaluation embarked upon. These changes were sufficiently entrenched that when the Cultural Revolution was launched in 1966 the implications for Liu Min Ying were different from those experienced by most other Chinese villages. The leadership were convinced of the correctness of the policies of Liu Shaoqi, despite his fall from grace during 1966 and consequently followed his line. According to Zhang Kuichang, farm manager and deputy Party secretary, while others were reading the works of Mao Zedong, the leaders of Liu Min Ying worked hard to improve the brigade economy through practical endeavour. While other neighbouring brigades were engaged in political movements, Liu Min Ying continued to emphasize economic advancement by improving the productivity of the land and building houses. The leaders of the brigade were criticized by the commune leadership for putting the economic improvement of Liu Min Ying above the official policy of building revolutionary proletarian consciousness. But the brigade was internally united and held its line.

In 1970, a new Party branch was set up under new *shuji* Zhang Zhanlin. Grain yields, at 2-300 *jin* per *mu* were still low and the soil suffered from saline-alkilinity. Thus, the leaders announced a Five-Year Plan, the only brigade in Zhang Zhi Ying commune to do so, calling on the villagers to 'work hard for a 1000 days, double the grain yield and change the economic situation' (Zhang, 1994). As a result, the saline-alkali soil was changed into high-yield square fields, which were double cropped, replacing sorghum and maize with rice as a main crop. The grain yield of Liu Min Ying in 1970 was 450,000 *jin,* by 1976 it had reached 1,000,000 *jin* allowing the brigade to add to its accumulation fund and invest in new irrigation equipment and farm machinery. As early as 1975, the *jiti* began to destroy the old mud huts and build brick houses in their place; public buildings, including a primary school and a public hall, were constructed. By the mid 1970s, at the end of the Cultural Revolution, Liu Min Ying was the most economically developed, highly mechanized and richest brigade in the commune. According to Zhang Kuichang, the favourable position of Liu Min Ying owed much to 'the correct leadership' of the brigade at that time, 'following very practical rules and policies and improving productivity, while nearby brigades followed blind policies.' In 1994, he said to assembled journalists:

> the two most important things in those days were the correct leadership of the Communist Party and the power of the *jiti*. Only we who had experienced these changes knew it most clearly (Zhang, 1994).

The leaders of Liu Min Ying did not wilfully flout official ideology : they simply considered that by ensuring there would be enough food to eat, the brigade would be a good place to live in. In 1990, when Deng Xiaoping's daughter, Deng Rong, visited Liu Min Ying, it was impressed upon her that its early economic successes, experienced during the Cultural Revolution, resulted from following the policies that Deng himself favoured at the time and which led to the criticism he (and Liu Shaoqi) experienced and to his fall from grace and enforced exile until 1975.

8(1) 3. Land Reform

Throughout the 1970s, Liu Min Ying maintained its vanguard position in terms of levels of mechanization and economic development. When the reforms of the late 1970s, were announced, therefore, it is unsurprising they were not embraced there with the same enthusiasm as in many other places. In particular, in common with many successful brigades where the leadership was respected, there was some reluctance to break up the *jiti* and replace it with the HRS.

While the leaders of Liu Min Ying were keen to avoid dividing the land amongst different

97

households, they wanted to divide it in another way - to produce different kinds of output. The reluctance to divide property was itself fuelled by the higher level of mechanization and the extensive base of publicly-owned property that the brigade had attained in comparison to nearby brigades which not only made the reforms less necessary but their carrying-out more difficult. According to Zhang Kuichang,

> Through a decade's hard struggle, we had got solid economic strength, good farming conditions, perfect leadership and high collective consciousness amongst the people. So if we then divided the fields, it was bound to destroy people's enthusiasm (Zhang, 1994).

Eventually, in 1982, 24 of the 25 brigades of Zhang Zhi Ying commune divided their property, but Liu Min Ying continued to resist. There was considerable pressure put on the brigade both by the commune and Beijing and the leadership faced considerable criticism for not doing so. Beijing Municipality sent out an investigating team to Liu Min Ying and subsequently Beijing radio reported critically that Liu Min Ying was cocking a snoop at official policy and was adopting its own responsibility system. But the unanimity of the leadership, the (eventually) unanimous support of the village and the higher level of farming encouraged continued resistance. A small minority of members of the brigade did initially want to adopt the HRS but after a series of full brigade meetings held to canvass opinion, they were persuaded to back the position of the leadership, which was forced, in turn, to placate the opponents of its strategy in Beijing by promising to adopt the HRS at a later date, even though it had no intention of doing so. The buying of time was successful: Liu Min Ying never adopted the HRS, the land and other communal property was never divided and the village remains *jiti* to this day.

It is paradoxical that Liu Min Ying should have been out of step with the official policy from Beijing in two such very different periods in recent Chinese history, i.e. during the Cultural Revolution on the one hand and during the early years of the reforms on the other. Solidarity with the officially disapproved-of lines of Deng Xiaoping during the former period did not lead to the village showing solidarity with Deng's favoured policies concerning land reform in the latter. This, according to Zhang Zhanlin, had nothing to do with deliberate contrariness, but more with the conservative nature of the leadership wishing to build pragmatically on past economic successes.

8(1) 4. Chinese Ecological Agriculture in Liu Min Ying

8(1).4.1 CEA: an introduction

1982 was a critical year for Liu Min Ying in more ways than one. At the same time as the land reforms were being considered, Professor Bian Yousheng of the Beijing Municipality Research Institute for Environmental Protection, was casting around for a village which would cooperate with his Institute's desire to construct an ecological agricultural project. Professor Bian first became aware of the possibility of the modern application of the principles of ecological agriculture to the Chinese countryside in the late 1970s as a result of programmes initiated in Europe and was keen to see the development of an experimental site for ecological agriculture in China. On his own initiative, although supported by the government and NEPA, Professor Bian visited a number of villages and state farms in the suburbs of Beijing in 1981, none of which were keen to participate in such an experiment, largely, he argues, because of lack of skills, uncertainty, lack of awareness of potential environmental gains and the perceived risks involved in disrupting current methods of production. Given the 'iron rice bowl' orthodoxy of the time, there was little incentive for state farms to alter production methods because the monetary rewards remained the same. Professor Bian further suggests that the negative attitudes of the village leaders were fuelled by suspicions that the experiments were designed more to boost the scientific acumen and status of the Institute's researchers rather than the villages' economies.

Liu Min Ying was spotted by accident, although from the start it did have one particularly attractive feature of its economy - it already used biogas, produced in a large number of small, household level pits. According to Bian Yousheng, that Liu Min Ying entertained the idea of the construction of an ecological agricultural project owes more to the powers of imagination and persuasion of Zhang Kuichang than to anything else. That boldness and imagination were necessary at that time is clear, given the underdevelopment of environmental understanding in China then. Before 1982, Zhang Kuichang was an ordinary peasant working the fields, his education having finished at secondary school. However he had already come to the leaders' notice because of his skill at writing slogans and his overall practical and intellectual abilities. Above all he was noted as a very practical man, a result, perhaps, of the lessons he learned during his six years in the PLA, some of which were spent on active service in Vietnam. In 1982, he was 31 years old, he joined the Communist Party and got promoted within the leadership of the village to become deputy *shuji* and Director of Farming.

It was, according to Bian Yousheng, the lead that Zhang Kuichang took in the critical months of 1982 in persuading others of the future benefits of CEA that sealed the fate of Liu Min Ying. When asked why he, Zhang Kuichang, had been enthusiastic to develop ecological agriculture, he says he was persuaded by the arguments of Bian Yousheng. Whatever else can be said, the energies and arguments of both men seem to have been critical factors in the initial acceptance of the project in the village. Today, both still have a deep respect for each other, despite their differing backgrounds, personalities and talents, clearly based on the critical relationship they forged in 1982.

It is fortunate that the decision to adopt CEA had to be made in the same year as the decision to divide the land or not. The village was already accustomed to village-wide meetings where people could express their opinions and Bian Yousheng addressed several of them. Had the decision to divide the land already been made, the likelihood of persuading villagers and their leaders into adopting ecological agriculture in 1982 would have been much reduced, given the level of intra-village cooperation necessary during the early years of the project. Meanwhile, the decision to adopt ecological agriculture helped postpone the higher authorities' efforts to force Liu Min Ying to abandon its commune and gave an incentive to the village leaders to maintain their line. In yet another way the two decisions were related: Liu Min Ying was, by the standards of neighbouring villages, rich, the result of successful economic development which was perceived by the villagers to be due, at least in part, to the superior level of social organisation there. They were thus unwilling to risk damaging that social organisation by dividing its communal property. However, it was its very wealth which presented the opportunity to construct CEA since the village had substantial collective savings not available to its neighbours.

In 1982 the collective's fixed assets amounted to 2m. yuan, output value was 590,000 yuan and the annual per capita income was 405 yuan. The brigade helped every household to purchase a colour television, washing machine, electric fan and solar energy stove and began to dig biogas pits and construct solar bathrooms. The slogan at the time was:

> Light a lamp without oil (biogas lamp), cook without any trouble (use biogas instead of stalks), have a bath without going outside of the yard (solar energy bathrooms), watch programmes in bed (television) (Zhang 1994).

From the start, the villagers had to be persuaded that the adoption of ecological agriculture would be a sound *economic* proposition: the villagers wanted to get rich. Though electric fans and television sets were already common, the villagers could not be persuaded by environmental claims on their own, if at all. The village itself did not suffer from any immediately obvious environmental problems, conscious environmental concern was in its infancy and no one had heard of ecological agriculture, not even the 43 year old *shuji* at the time, Zhang Zhanlin. Zhang Zhanlin himself recalls that he was persuaded to recommend adoption of ecological agriculture by the prospect of it providing *long-run* prosperity: while other villages were beginning the headlong rush into manufacturing industry lured on by short-run profit, he felt that rapidly rising costs might choke off that profitability in the longer term. He was also worried, along with other leaders and villagers that grain monoculture could not sustain long-run prosperity. Zhang Zhanlin suggests that he himself was ultimately the critical decision-maker. He argues that the villagers needed a great deal of persuading, that Bian Yousheng played a crucial role in so doing but that the decisions that were finally agreed upon could not have been carried through without (his own) strong leadership. All the principal players agree that resolute leadership was decisive at this time. The village was promised substantial technical help from the Beijing Municipal Research Institute of Environmental Protection in developing the project and a subsidy of 16, 000 RMB from the same source. Otherwise, it had to rely financially on its own resources and take out loans on its own account. That no other village in the suburbs of Beijing was prepared to contemplate adopting ecological agriculture in these circumstances emphasises the boldness of the village and its leaders to do so.

Thus, in mid-1982, the decision was made: the preparation for the project began in November and the project itself in 1983. Responsibilities were shared between Beijing Municipal Research Institute and the village: the former was responsible for the overall planning, research work and design of the project, the latter for its construction. The initial plans divided the project into three stages, 1983 capital construction, 1984-5 adjustment, 1986 completion and perfection. Over the first three years, up to 40 engineers from the Beijing Institute lived and worked in Liu Min Ying, staying in the village even during Spring Festival.

While the timetable was adhered to early on, it is clear that 'adjustment, completion and perfection' did not stop in 1986. Those processes continued throughout the 1980s and inevitably, in some senses, are continuing today.

8(1) 4.2 CEA construction

The construction of CEA in Liu Min Ying involved a radical transformation of the structure of the village economy. The main elements of the project, all clearly interrelated, were: (1) radical shifts in forms of production of the village, (2) development of new energy sources, (3) the utilization and recycling of organic wastes and (4) research into methods of improving soil fertility (Bian 1988, pp.384-397).

In 1982, the economy of Liu Min Ying was uni-dimensional, based as it was, in common with most parts of the Chinese countryside which was still taking 'grain as the key', upon the production of wheat and rice. The total value of output in that year was 693,000 RMB, the proportion of output attributable to different elements of production being:

crops (wheat and rice):	78.4 per cent	
poultry and animal husbandry:	6.0 per cent	
industry and side-line production:	13.0 per cent	
forestry:	0.3 per cent	
fisheries:	0 per cent	(Bian, 1988, p.385)

Thus, the economy was overwhelmingly dependent upon grain. Bian Yousheng and his team argued that the structure of the economy was unbalanced, leading to waste of resources and high economic and environmental cost. One obvious waste was underemployment, in effect, unemployment of 70 of the 400 workforce in 1982, a problem which was becoming perceptibly worse as the years rolled by. The production structure, based on grain, was simply unable to deliver sufficient employment for its people. But there were many other sources of waste. The productivity of labour and land in the primary sector, after the improvements made in the 1970s, remained static. Meanwhile, since secondary production was so undeveloped, the outputs of the primary sector were not being used effectively within the village economy. And neither were its by-products: crop residues were frequently wastefully burned. At the same time, there was considerable use of chemical fertilisers on the land, by 1982 to the tune of 150 kilograms per *mu* (Bian, 1988, p.385). The use of chemical fertilisers poses considerable environmental risks but for the peasants of Liu Min Ying, it entailed an ever increasing direct cost of production. The negligible forest cover not only involved predictable degradation of soil quality through loss of biomass and wind erosion but led to increased difficulties and cost of obtaining firewood.

Thus, one of the primary aims of the project was to diversify the structure of production, away from one wastefully dependent upon crops, towards a more balanced one where factors of production could be more efficiently employed and the outputs and by-products of one sector could be usefully employed as inputs in others. This involved a conscious attempt to develop 'environmentally friendly' sideline industries appropriate to the outputs and needs of other sectors of the village economy, expansion of poultry raising and animal husbandry, and investment in forestry and fisheries. With regard to the former, the village built several factories, including a fodder processing factory, a flour processing factory, a soft-drinks factory and an abatoir. With regard to the second, a chicken farm, a duck farm and a pig farm were built, trees were planted and fishponds dug. As a result, within three years the economy had been substantially transformed. By 1985, the value of output has risen fourfold over 1982 to 2.8m yuan and the proportions of that output attributable to different sectors of the economy were: industry and sideline production: 55.3 per cent (up from 13 per cent) crops (wheat and rice): 26.1 per cent (down from 78.4 per cent), poultry and animal husbandry: 12.4 per cent (up from 6 per cent), forestry: 0.8 per cent (up from 0.3 per cent) fisheries: 0.5 per cent (up from zero) (Bian,1988,pp.385-6).

Since 1985, further significant transformation has taken place. Liu Min Ying has continued to embrace rural industrialization as more sideline enterprises have been established and animal husbandry and poultry raising have continued to expand as dependence on crops has diminished. Diversification has thus succeeded in the first key object of the ecological agricultural project: the transformation of the uni-dimensional production structure extant in 1982.

As explained in Chapter 2, while technological change has made modern agriculture increasingly efficient in terms of the productivity of land and labour, it has become ever more inefficient in its use of energy. As a result a key aim of CEA was to pose an alternative to the 'petroleum agriculture' of the West and develop new, more environmentally-friendly energy sources. From the start biogas was central to the construction of CEA in Liu Min Ying.

As noted earlier, by 1982 Liu Min Ying already possessed biogas, having built 158 small household level biogas pits of approximately 8-10 cu.m capacity in the 1970's during China's first major biogas drive. At the start of ecological agricultural construction in 1983, a further 12 household level pits were built. It was felt that the earlier pits had many technical problems: the rate at which they produced biogas was low and their useful life (3-4 months a year) short. One of the key elements of the project was to research ways in which these problems could be overcome. The research programme concluded that the rate of biogas production could be significantly increased if there was careful handling and mixing of the inputs, recommending that the ratio of (wet) manure to (dry) stalks should be 3:2, that the dry materials should be cut into small pieces, piled up and soaked with high quality stimulating liquid, adding fresh manure or liquid ammonium carbonate and that stirring devices in the pits should be constructed. In order to prolong the life of the pits themselves, it was recommended that ditches should be dug using plastic shields for heat preservation in winter and that the biogas materials should be regularly changed. Experimentation subsequently suggested that, by using the recommended methods, it was possible to increase the rate of biogas production by up to 50 per cent (from 0.08cu.m of biogas per cu.m. of pit to between 0.10 and 0.12 cu.m) and that the working life of a pit could be prolonged from 3-4 months to 6-8 months in a year. One experimental pit worked for 291 days, providing enough energy for cooking 3 meals a day and lighting at night.

Of the 170 pits existing in 1986, 30 per cent were performing to the improved levels (Bian, 1988, pp.387-8).

Experimentation with solar energy also took place. In the early years of the construction 180 solar cookers were constructed, each capable of being used for over 2,700 hours in any one year and 165 solar water heaters with a capacity of 1.5cu.m. were installed, usable for 171 days a year, as were 38 solar heated rooms, usable for 96 days a year. Bian Yousheng estimated in 1986 that the total economic benefit enjoyed by the villages over the previous 3 years from biogas and solar power amounted to over 98,000 yuan and that a saving of 362 tons of coal had been made (Bian, 1988, pp.388-9).

Central to the construction of the project was the establishment of a virtuous ecological cycle whereby waste outputs of one activity were used productively as inputs for others. Thus its construction depended on the diversification of production to include animal husbandry and fish farming on the one hand and further development of biogas on the other. Stalks and plant residues were fed to the animals, including chickens, ducks, pigs and cows whose night soil, combined with dry material, comprised the input for the production of biogas. The biogas waste slurry was then either returned to the fields, used as fertiliser for mushroom cultivation or deposited into the fishponds as food for the fish. The bottoms of the fishponds were then dredged and the mud recycled back to fertilise the fields. Mini-cycles existed within the larger framework, the most important being the feeding of pigs with a mixture of ordinary and anaerobic fodder with fermented chicken droppings.

Research into methods of improving soil fertility represented the last key elements of the construction phase of CEA. The research discovered the content of the organic nutrients in the soil and various other indices of soil quality, such as its respiration intensity, ammonium intensity, the total number of bacteria and the number of natural nitrogen-fixing bacteria. Results of the research suggested that barnyard manure was the most efficient fertiliser and that its application put the physical, chemical and biological properties of the soil at their best. Biogas manure was seen to be the second best manure. While chemical fertilisers could increase grain production, the research suggested it had negative effects on the physical, chemical and biological properties of the soil.

Research on the land of Liu Min Ying concluded that the soil contained sufficient nitrogen but lacked potassium and phosphorus and that to improve the fertility of the soil, a variety of measures should be taken, including returning wheat stocks back to the fields after fermentation in the biogas pits, rather than burning them. It was recommended that the proportion of organic fertiliser (from barnyard and biogas slurry) to inorganic fertiliser should be significantly raised with the latter being applied in a nitrogen, phosphorus and potassium ratio of 3:2:1 (Bian,1988,pp.393-4).

Various other recommendations were made concerning optimum ploughing techniques and the advisability of growing green manure in parts of the larger grain fields. Use of chemical fertilisers and pesticides which were harmful to the activity of microbes in the soil and which had, as a result, reduced the number of nitrogen fixing bacteria were to be, as far as possible, phased out.

Bian (1988,pp.417-436) evaluated the results of the first four years of ecological agricultural construction in terms of economic, ecological and social benefits. As discussed earlier, the decision to adopt CEA in Liu Min Ying was based on the understanding that economic and social benefits would materialise. Bian Yousheng and his team produced, for the years 1982-85, a range of economic indicators suggesting considerable economic benefits. In terms of gross value, output increased very considerably over the three year period, (see table 1 below). 1982 represents the last year of traditional agriculture and 1985 the first full year of CEA.

Table 1	Gross output value ('000 yuan)	
	1982	1985
grain	538.2	732.6
forestry/fruit	13.8	22.7
animal husbandry	41.4	347.6
industrial sideline production	88.7	1550.0
fisheries	-	14.7
others (commerce and construction)	7.8	136.0
TOTAL	**687.9**	**2803.6**

(Bian 1988, p.419)

The 1985 figure represents an increase of 306 per cent on the 1982 figure, an average annual increase over the 3 years of 64 per cent. This contrasts with a figure of 25 per cent for the average annual increase in production in the suburbs of Beijing and 12 per cent for the country as a whole. In that the numbers of workers and size of the population remained almost constant, this 64 per cent annual increase in production represented a similar increase in output per capita and output per worker.

In terms of the physical output of goods, similar increases took place, see table 2.

Table 2	Output (kilograms)	
	1982	1985
grain	935,000	788,000
pork	7,000	15,400
milk	0	80,000
fresh eggs	0	75,000
beef	400	15,000
chicken meat	0	7,500
fish products	0	8,500
vegetables	50,000	600,000
preserved eggs	0	5,000
drinks (10,000 bottles)	0	150

(Bian 1988, p.424)

In terms of per capita income per head, similar rises took place. In 1982, per capita income produced was 504 yuan, per capita income distributed was 405 yuan. In 1985, these figures had increased to 1250 yuan and 1000 yuan respectively, more than doubling distributed income per head in three years. The initial construction of CEA allowed a significant economic advance to be made in the early years making Liu Min Ying substantially richer than many other Chinese villages. While by 1985 per capita distributed income was 1000 yuan, the average figure for the Chinese countryside was 369 yuan. The average villager of Liu Min Ying was, in terms of income per head, almost 3 times richer than the average Chinese peasant. Meanwhile, the underemployment of villagers (70 were unemployed in 1982) had been transformed into a shortage of labour to the tune of 13 or 14 a year.

There was an improvement in the ratio of organic to inorganic nitrogen in Liu Min Ying.The study showed that the higher that ratio, the more efficient the productive outcome. When the ratio of organic to inorganic fertiliser was 0.5:1, the consumption of nitrogen was 2.3 kilos per 50 kgs of rice, when the ratio was raised to 0.922:1, the consumption of nitrogen fell to 1.6 kilos. per 50 kgs. of rice. In Liu Min Ying, the proportion of organic to inorganic fertiliser rose annually with the construction of ecological agriculture. In 1982 the ratio was 0.267: 1, in 1983 0.32 : 1, in 1984 0.70 : 1, and in 1985 0.83 : 1 (Bian, 1988, p.427). Accompanying this increase there was a doubling of the output:input ratio of nitrogen elements, from 0.25:1 in 1982 to O.5:1 in 1985 (Bian, 1988, p.427)

There were measurable improvements in the condition of the soil owing to the greater use of organic fertiliser, the utilization of organic waste matter and the growing and utilization of green manure. For example, the mean percentage of organic matter in the soil in 1982 was 1.22 per cent, by 1985 this had risen to 1.65 per cent. This meant that the application of chemical fertilisers was considerably reduced from 300,000 kilograms in 1982 to 150,000 kilograms in 1985, significantly reducing farm costs (Bian, 1988, p.429).

There was an extension of forest cover. In 1982, the percentage of Liu Min Ying with tree cover was 8.7 per cent. In the period 1983-5, 12,000 trees were planted with the result that by 1985, forest cover had reached 12 per cent, with trees lining the roads, fields and farm surrounds (Bian, 1988, p.430).

There was an improvement in the general aesthetic environment and in the sanitary and health conditions in Liu Min Ying. By 1985, seven flower beds and 25 *mu* of lawns had been planted in the village, 24 flower nurseries had been established and 2,000 fresh flowers planted along the roadside, brightening up and 'greening' the general look of the village. Meanwhile, the feeding of human and animal excrement into biogas production led to the destruction of pathogenic bacteria, significantly reducing the (previously high) incidence of intestinal diseases and improving the general health of farmers. The use of biogas and solar energy for domestic cooking, lighting and heating purposes, in the estimation of Bian Yousheng, had in the 3 years 1983-5, saved the consumption of 362 tons of coal (1988, p.389), saving 98,800 yuan, and implying an obvious 'win-win' process with regard to the economy and the environment, given the reduction in the need for coal-burning.

Though the enforcement of the State family planning policy was not directly part of the construction of ecological agriculture, the two processes were interlinked. The one-child policy, in the early years, was largely adhered to, leading to a reduction in the natural growth rate of the village population from 1.69 per cent p.a. to 0.69 per cent p.a. With regard to other social benefits Bian Yousheng argues that the construction of CEA had, by 1986, made achievements in 6 'social' aspects. It had provided a beneficial experience for building 'socialism with Chinese characteristics' (*you Zhongguode tesi shehui zhuyi*), 'welcomed by the workers' (Bian, 1988 p.433), helping to strengthen and modernise Chinese agriculture and the economy generally, shown interconnections between spiritual and physical civilisation, shown solutions to the problem of rural unemployment, illustrated the importance of developing farmland for the rural economy, shown that ecological agriculture was able to play a full part in stimulating farm and sideline production and thus in activating and enriching the marketplace, and fostered a scientific culture in the countryside.

8(1) 5. Political Economy in the 1980s and 1990s

8(1) 5.1 Political structures

Despite enormous economic change, there remains considerable continuity in institutional and personnel terms. The village continues to be run on *jiti* lines, headed by a six-member committee (all CP members), nominated and chosen by the villagers, who hold office with the blessing of senior cadres in the township of Zhang Zhi Ying. The six include the leader of the village, *shuji* Zhang Zhanlin, director of farming and deputy *shuji* Zhang Kuichang and Professor Bian Yousheng, nominated by the Beijing authorities. Beneath are three tiers of administration. At the lowest tier are the fields and sideline industries. The collective works through a contract system whereby the farm director, Zhang Kuicheng initially lays down quotas which must be fulfilled for units at lower administrative tiers, as do the managers lower down the line. Those enterprises which oversubscribe their contracts are able to sell their surpluses on the open market in Beijing or Tianjin. The revenues received are then distributed in the following ratio: 50 per cent is kept by the individuals themselves, 30 per cent is reinvested in the enterprise and 20 per cent is returned to the collective. Taxes are then paid to the township from these latter revenues. In 1995, a total of 120,000 yuan was paid by Liu Min Ying to Zhang Zhi Yang in the form of agricultural tax *(nongye shui)* and 400,000 yuan from industry. Other obligations on Liu Min Ying towards superior political authorities are slight. Though it must sell grain, pigs and eggs to the State at low (ie below market) prices, the quantities are very small: in the case of grain, only 100 kilograms, out of a total output of grain of 2m. kilograms.

All workers earn roughly the same in different enterprises, roughly 800 yuan per month in 1997, although agricultural workers earn slightly more. Workers are allocated to particular activities by Zhang Kuicheng as far as possible on the basis of individual preferences, skills, aptitudes and abilities: some activities are more popular than others and ultimately the decision rests with Zhang Kuicheng. In the collective, while subject to periodic nomination from below and confirmation from above, the leadership still yields just about absolute power.

A major development in Liu Min Ying, in common with most of rural China, has been the expansion of sideline industries, although this has taken place slowly, with care that what expansion has occurred should not pollute the environment. The development of township and village enterprises (TVEs) in the early 1980s emphasized links with agricultural processes, with the establishment of the abatoire, fodder factory and flour processing factory. In the late 80s developments included the soft drinks factory and a hardware factory making small machine parts. Meanwhile, in 1991, a factory making dried flowers was added to the productive capacity of the village (although it closed in early 1996 for lack of profit).

In early 1993, major new initiatives were undertaken: the construction of a factory making porridge *(babao zhou)* as a partly owned joint venture with a Taiwanese company was begun although within two months the Taiwanese had pulled out and a Chinese company, Paren Tourism Products Company Ltd, signed a contract and took over the factory to produce quilted sleeping bags, reopening for production in October 1994, becoming one of the three most important factories in Liu Min Ying by 1997. In September 1993, construction began on a wholly foreign funded canning factory, financed to the tune of 2m yuan by a company from the USA, attracted to Liu Min Ying by the environmentally-friendly reputation the village had gained outside China and in the same year plans were hatched to go into a joint-venture operation with a Japanese company to grow 'green' vegetables to be sold and marketed in Japan. In both cases these projects had fallen through by 1995, although the latter was reactivated in 1997. In 1994 a joint-venture contract was signed with Bei Qimo radiator company (a subsidiary of the Beijing Automobile Company) to make car radiators and air-conditioners, and the factory so to do, originally envisaged as the canning factory, opened in early 1995. The contract involved Liu Min Ying providing the site for the factory and its construction, while Bei Qimo Radiator Company provided the initial finance, management and technology. By 1996, the radiator factory was much the most important in terms of employment and output in the village (remaining so in 1997), employing 200 workers, 100 of whom were being bussed daily from Beijing. In 1996, profits to Liu Min Ying from the factory topped 1.5m. yuan. The total value of industrial output in Liu Min Ying in 1996 was 80m. yuan, almost double the income from animal husbandry and agriculture combined.

Decision-making with regard to this potential influx of foreign capital, as with all decisions pertaining to the introduction of new economic activity, remains with the village leaders with support from the township. However, given the recent experiences with Taiwanese and Japanese companies, the leaders are less interested in foreign money than they once were. In 1994 Zhang Kuichang said they now preferred to deal with Chinese companies, 'the Taiwanese and Japanese only want to cheat the local people.' However, this did not stop the village from signing a joint-venture contract with a Japanese company in 1996 to open a dumpling *(jiaozi)* factory the following year, the Japanese providing an initial 750.000 yuan for its construction nor from reactivating the Japanese joint-venture operation involving 'green' vegetables.

Indeed, there has recently been a significant increase in the importance of vegetables, fruits and gardening with a related decline in the importance of crop growing. In 1994, 80 *mu* of arable land was sacrificed and in 1995 a further 100 *mu,* in order to grow 'green' vegetables, mainly tomatoes, cucumbers, onions and aubergines, in polythene greenhouses, now stretching over 350 *mu* of previously arable land. A further 120 *mu* are given over to fruit growing and 130 *mu* to gardening. By 1996, vegetable growing employed more workers than were employed in the growing of crops (120 as opposed to 20) and Liu Min Ying produced 6m. kilograms of vegetables compared to only 1m. kilograms of grain. Increasingly more farm output is classified as 'green': in January 1996, Liu Min Ying was granted the status of 'Green Food Base' by the Beijing Green Food Centre (operating under the Ministry of Agriculture) which tests the village's vegetables, fruits, meat and crops. Wholesalers come to the village who then sell it on into elite markets in Beijing at a 10 per cent or so premium. Negligible amounts of chemical fertilisers are used in the growing of fruits and vegetables which are fertilised almost exclusively from biogas slurry. Since 1995 1.1m yuan has been spent on new agricultural machinery and irrigation sprinklers for all fields. The total value of agricultural output in 1996 in Liu Min Ying was 8m. yuan.

Animal husbandry has been further developed and in 1994 there were 4 animal farms: a chicken farm with 100,000 laying chickens, a duck farm with 200,000 ducks, a pig farm with 500 pigs and a dairy with 100 cows. In all, Liu Min Ying produces 1.3m kilograms of eggs, 300, 000 chickens and ducks, 90,000 kilograms of beef and pork and 200,000 kilograms of milk. the total value of the output of the animal husbandry sector in 1996 was 36m. yuan. In 1995, a new roasted duck factory was opened, the ducks from the village's own farm being slaughtered in a new slaughter house, then roasted and sealed in polythene bags to be marketed for sale throughout the country.

In terms of employment, the 70-odd workers under-employed before the construction of CEA have disappeared and the problem of increasing rural unemployment, so evident and serious across China in the 1990s, has not taken place. Indeed, a shortage of labour has meant that by 1997 there were over 500 migratory workers living in dormitories in Liu Min Ying, mostly young girls from Sichuan, some from Inner Mongolia and the rest from Henan province, while a few villagers from neighbouring villages were also working in Liu Min Ying. Since the recent expansion of the radiator factory and animal husbandry a hundred or so workers get bussed into Liu Min Ying from Beijing daily. Meanwhile, long-term movement out of the village by the young in search of jobs has not taken place.The farm output of the village in 1992 was as follows: (1985 figures in brackets) grain crops: 1,200.000 Kgs (788,000 kgs), despite a reduction of cultivated area from 1,600 *mu* to 1,200 *mu.*, vegetables: 3,000,000 kgs. (600,000 kgs.) eggs: 1,300,000 kgs. (75,000 kgs.), milk: 200,000 kgs. (80.000 kgs.) fish products: 20,000 kgs. (8,500 kgs.), meat (pork,beef and poultry): 80,000 kgs. (37,900 kgs.), fruit: 70,000 kgs, (negligible). By 1996, the grain output was down to 1m kgs., the output of vegetables up to 6m.kgs.

Total village income in 1992 was 2.4m. yuan while output per worker was 3,800 yuan and distributed income per capita 1,900 yuan. In 1995, this latter figure was 3500 yuan, compared with a figure of 405 yuan per head in 1982, representing a growth rate approaching 25 per cent p.a. for the period, a rate on average twice that experienced by the Chinese as a whole over the same period and only bettered by a very few in privileged locations or fortunate market positions. Moreover, in that annual income per head in Daxing County in 1995 was 1850 yuan the statistics suggest that, despite fast rates of growth experienced in the Beijing municipality in the 1980s the villagers of Liu Min Ying were, on average, at least twice as well off as their rural neighbours in the early 1990s. Thus the economic lead that Liu Min Ying had over its neighbours as it entered the 1980s was maintained relatively and absolutely 15 years later. By 1996, income per head had risen to 4500 yuan per annum. According to Zhang Kuichang in 1996 only satellite villages much closer to Beijing where money was being made by buying and selling property for residential purposes could boast anything higher.

In 1996, a new project was decided upon and initiated: the development of eco-tourism in conjunction with a Hong Kong company which put 20m. yuan into the project in the first year of its construction and has promised 80m. more. The village involves the construction of villas on ecologically sound principles, including solar energy, with living, dining and entertainment facilities, to attract tourists to Liu Min Ying for recreational purposes. A model of the Great Wall was built in 1997 and a lake was being constructed. The development of eco-tourism is just one of a number of new projects within a now expanding service sector: in the last 3 years two private grocery shops and three private restaurants have been initiated (under licence from the *jiti)*, while the *jiti* itself runs a taxi company with 30 taxis and a driving school.

8(1) 5.3 Social developments

Higher indicators of economic performance have translated into considerable improvements in material standards of living in Liu Min Ying since 1982. Much improved social infrastructure including housing and the opportunities to enjoy now conventional trappings of material comfort inside the home - colour televisions, video-recorders and washing machines - are the most obvious manifestations of that. In 1993 the average household owned 16 such items. Meanwhile, the monotonous diet of steam cornflower of a generation ago has been replaced by a varied one including rice, meat, vegetables, eggs and fruits.

The improvements in material standards of living are not limited to increased income per head and greater availability of consumer durables. There has been a comprehensive housing programme, so that most families in Liu Min Ying now live in two-storey and some in three-storey houses. The total residential space in 1993 was 45,000 m^2, providing 50m^2 for each villager, approximately five times more than was available 25 years previously. Housing continues to be improved every year.

New asphalted roads have been constructed and maintained, including roads linking Liu Min Ying to the Beijing-Tianjin expressway. A range of other public facilities have been built. While in 1982 the only such facility was a clinic which had been built in 1975, during the 1980's new facilities were added, including a primary school in 1985, a kindergarten in 1986, a leisure facility (with ping-pong, mah jongg etc.), an improved canteen, administrative offices and reception building all in 1987 and in 1993, a new village shop selling groceries and household goods. A very attractive peoples' park covering 10 *mu*, with gazebo, ponds, bridges, flower beds, grass and a huge smiling Buddha was completed in 1995 and much improved by

1997. In 1995, a brand new and very impressive education building of 1,700 m^2 was opened to house the primary school, kindergarten (to which all children from Liu Min Ying go free of charge) and adult education unit as was an auditorium which seats 1000 people and which is now used for conferences, party meetings and entertainments, including concerts and cinema. And in 1996, a brand new civic centre costing 5m. yuan and containing a very impressive complex of administrative buildings including a guest house was constructed and opened, all paid for from the profits of the village industries. In terms of social infrastructure, therefore, by the mid 1990s Liu Min Ying had become barely recognisable from even five years before. In 1994, sixty households had phones using 9 direct telephone lines to Beijing, only 2 had existed a year earlier. According to Zhang Zhenying, a 55 year old female villager, in whose home I was given demonstrations of cooking with biogas and boiling water using solar energy, explained that things had improved immeasurably over the years. 'There are as many things to do here as in a town'.

In terms of social welfare, by 1993, a health insurance scheme existed whereby all families were given 10 yuan per year to cover treatment at the clinic: if a worker became ill or injured as a result of work, they were recompensed fully for lost wages and bonuses by the collective, if a women was pregnant or workers became ill for non-work reasons they were recompensed for lost wages, although not for lost bonuses. Retired workers remained the responsibilities of their families; each retired worker was provided with 25 yuan per month from collective funds.

One further social benefit, according to Bian Yousheng, is the example Liu Min Ying has set in showing that rural unemployment and outward migration are not inevitable in China in the 1990s. For Bian Yousheng, the 'social benefit' is not limited to the enjoyment of the villagers themselves but is extended to the locality and the nation.

8(1) 5.4 Environmental developments

The UNEP 'Global 500' award scheme was inaugurated in 1987 with the intention of rewarding individuals and organisations throughout both the developed and developing worlds with recognition for environmental achievement. In that year, the National Environmental Protection Agency in China made a number of nominations of which three were accepted by the UNEP as deserving of the opportunity of being among the first recipients of the award. Alongside China Environmental News and The Great Green Wall Group, Zhang Zhanlin was so recognised. His citation was as follows:

> As head of Liu Min Ying production brigade, he popularised biogas and solar energy in his community and initiated ecological farming using agricultural wastes for energy and fertiliser. His work is a model for renewable energy development.(UNEP 1995)

In that the relevant literature (e.g.Qu 1991, NEPA 1992) always refers to Liu Min Ying village as the recipient of the award, it was surprising (to me) to find that its *shuji* was, in fact, the citee. Bian Yousheng explains the fact by suggesting that NEPA was keen to see individuals as well as groups be awarded the honour and since the other recipients in 1987 were groups, it fell to Zhang Zhanlin to receive the award as an individual. Given the importance of status in China, it would clearly not have been awarded to any other individual in the village, however strong his or her claims might have been. In that Liu Min Ying as a village (rather than its *shuji*) is always referred to as the recipient of the award - even by NEPA - however, suggests that it was the village that was essentially being honoured. Liu Min Ying thus became the first village that had adopted CEA to join the UNEP's Global 500 roll of honour for environmental achievement. As such it earned a local and national reputation which it has strived to maintain since.

Thus environmental developments have been consolidated: in particular, the virtuous ecological circle has been expanded and deepened as animal husbandry and more recently vegetable growing has played an increasing role in the economic activity of the village. But the major development since 1987 has been regarding biogas. The construction of a village level 100 m^3 high temperature fermentation biogas digester was completed in 1993 and came on stream in the same year. For a variety of reasons, despite the improved design and methods of the new household level biogas pits built in the early 1980s, they still exhibited technical problems. While the UNDP and the Beijing Research Institute took an active interest in the process, the final decision to do so was made by the village leaders alone and the 700,000 yuan capital cost was financed entirely out of village funds, generously boosted by a gift from a Canadian for the purpose. In 1995 it provided power for cooking to 200 of the 240 households in the village, a minority clinging to individual biogas pits on grounds of cost. Bian Yousheng estimated in 1992 that the household level biogas pits saved 120 tons a year, which would imply that if only half the surrounding villages had followed Liu Min Ying's lead, the total energy demands of the rural Beijing municipality would be reduced by one third. According to Zhang Guanghui, deputy farm director in charge of biogas operations in Liu Min Ying in 1994, the new biogas digester saved 420 tons of coal, which at a cost of 190 yuan per ton, represented a total monetary saving of around 80,000 yuan per annum for Liu Min Ying alone. If other villages in Beijing municipality were to follow *this* example, then it is clear that total energy demand in the countryside would be seriously reduced.

As far as the costs and benefits of the first community biogas digester to Liu Min Ying are concerned, the costs (of 3 or 4 workers' salaries and the interest on a loan of 600,000 yuan) have been significantly outweighed by the benefits accruing in the shape (a) economies in the use of coal, (b) reduced consumption of chemical fertiliser and (c) the environmental benefits associated with greater availability and application of enriched organic fertiliser and 'clean' energy for household cooking. The village leadership remains optimistic about the prospects for biogas because of its economic success: each household pays 15 yuan a month for biogas, helping to provide a return on the initial investment while households benefit by paying less than they would otherwise do for canister gas (up to 40 yuan a month).

In 1995, plans were initiated to built a second 200 m^3 biogas digester, construction began in September 1996 and the digester was due to begin operation in October 1997. The expansion of investment in biogas reflects not only excess production of animal excrement but faith in the technology amongst Liu Min Ying's leaders and a grant of 1m. yuan from the Beijing government for the purpose. The extra biogas produced will provide household heating as well as lighting. Zhang Kuichang emphasized that with the exception of road construction, Liu Min Ying had not benefited from financial help from outside before.

Associated with the improvements in housing, the spread of running water and the increased use of biogas, sanitation levels have improved. Meanwhile, in 1997, the village embarked on an ambitious programme of providing a solar-powered bathroom unit for each household based on a revolutionary new design produced by Qinghua University, providing hot water in winter as well as summer.

Other environmental improvements consolidated since 1987 include the continued increase in the productivity of the soil as the use of organic fertilisers, whether from barnyard manure, biogas residue or duck pond mud, has substantially replaced chemical fertiliser while the crossplanting of the fields with trees and crops has enhanced the soil quality while reducing soil erosion from the wind. While in 1982 the average organic matter in the soil was 1 per cent and the highest 1.2 per cent, the average had risen in 1992 to 2 per cent and the highest to 3.5 per cent.The virtual elimination of chemical fertilisers (in 1982, the village used 250,000 kilograms of chemical fertiliser, in 1992, it used 8,000 kilograms) and pesticides has not only reduced the bills of farmers but the long run contamination of the land and watercourses while improving the quality of the soil and crops. Meanwhile, forestry has continued to expand: in 1982 forest cover was 6.0 per cent, by 1994 it had risen to 20 per cent. By 1996, Zhang Kuichang claimed it had risen to 23 per cent. In 1996 there were 2 hectares of fishponds.

With the increased use of animal waste there are fewer disease-carrying insects and the general level of health in the village has continued to improve with fewer villagers falling ill from intestinal diseases. Aesthetically, the environmental is enhanced as more trees are grown and flowers planted. The roads are swept daily. To an observer accustomed to associating aesthetic beauty and charm with an archetypal English village, with church and pub around the village green, Liu Min Ying is decidedly neither aesthetically beautiful nor charming. But to the observer who is accustomed to the abounding pollution and environmental decay existing in most of the villages currently undergoing rapid industrialisation in China, Liu Min Ying is clean, relatively green and orderly. In 1982, an application by an enterprise engaged in electroplating to set up in Liu Min Ying was turned down by the village council on environmental grounds (electroplating being an industry frequently leading to serious water pollution) and the decision 'despite one or two dissenters' received general support.

A further aspect of Liu Min Ying's environmental development has been its birth control policy. Despite initial bland assurances of the village's strong record of implementing the national one-child policy, it is clear that it needed considerable reinforcement before it was successfully enforced. According to Li Xueming, responsible for the one-child policy in 1993, it was initially put into effect by fining those who had a second child. However, the disincentive effect of this was reduced as the village got richer and more people could afford the fine. Eventually, enforcement involved sterner measures: a responsible leader was appointed who kept records of each woman, monitored all pregnancies and who had the responsibility of ensuring, by education or coercion, that women who had already given birth did not do so again. Big character slogans reinforce the message and abortion is a significant part of the family planning process. The new policy has been a great deal more effective than the old one and the leaders claim almost 100 per cent success in upholding it in recent years. However, Li Xueming stressed that as people had got richer they now considered the cost of having, rather than not having, children.

However, one problem , in common with much of rural China, has been the loss of arable land to building. In 1982, the total amount of arable land devoted to crops in Liu Min Ying was 1640 *mu*. In 1993, this figure had fallen to 1,200 *mu,* and by 1996 to 1000 *mu,* a loss of over 40 per cent in 14 years.This loss has resulted primarily from the construction of greenhouses for vegetables and from fruit growing and gardening although the construction of the chicken and pig farms, housing, public buildings and factories has also been at some expense of arable land. In 1997, the eco-tourist village and 'Great Wall' were being constructed on erstwhile arable land. Bian Yousheng points out that despite the loss between 1982 and 1992, overall output of grain grew because of the significantly enhanced productivity of the land, up from 650 to 1005 kgs of grain per *mu*. This process cannot continue indefinitely, however, without serious consequences both for the economy and the environment.

8(1) 6. Conclusions

Liu Min Ying is often referred to in the literature as China's first ecological agricultural village. To the extent that it was the first village consciously to construct an ecological agricultural system and the first to win the UNEP 'Global 500' Award, the description is accurate. Moreover to the extent that in the mid-1990's, incomes per head in Liu Min Ying remained higher than most other eco-villages I visited, without any obvious compromises having been made on the ecological principles initially adopted, it would suggest that the description may not be entirely unreasonable in other respects.

Liu Min Ying has been since 1982 an archetypal 'model' village. Bian Yousheng argues the primary importance of Liu Min Ying has been to set an example to other villages, to show that it is 'possible to succeed' in making economic progress without degrading or polluting the environment. It regularly attracts visitors from all over the country and from abroad; former Premier Hua Guofeng, senior leader Deng Xiaoping's daughter, Deng Rong, Premier Li Peng, former, now disgraced Beijing Party chief *shuji* Chen Xitong, current State Councillor with responsibility for environmental protection Song Jian, all have visited Liu Min Ying alongside 180,000 of their compatriots, as have Robert Mugabe and Imelda Marcos, among an estimated 8,000 foreigners from 120 countries, an estimate made to me by Zhang Kuichang before the UN Women's Conference in September 1995, when Liu Min Ying became a regular destination for overseas female guests eager to escape the confines of the conference halls in Beijing and Huairou. It is certainly a well-known village within the ecological agricultural community and is perceived by leaders of other eco-villages to be a worthy example, despite their jealousy of its many perceived advantages. It has a good reputation in Daxing County and is, according to Wang Yuling, vice Head of the Daxing County Environmental Protection Office in 1993, used by her office as a model of rural economic development.

While Liu Min Ying *is* a model village, it is by no means an ostentatious one: in 1992, it was exceedingly difficult to find (I eventually arrived on the back of a horse-drawn cart after a five hour journey from Beijing) and there were no obvious outward signs of its 'model' status. By 1993, arches had been built at either end of the village welcoming visitors to Liu Min Ying in Chinese, with an English translation added to the back gate in 1994. However, there were no flashy motor cars to show visitors around in, as in some eco-villages: indeed, when I arrived with a team from China Central Television (CCTV) to make a documentary about Liu Min Ying in July 1995, Zhang Kuichang showed us around riding a rusty old bicycle in front of the CCTV car. In my frequent visits to Liu Min Ying, even with the CCTV crew, there were no lavish banquets. Though Zhang Zhanlin *is* driven back and forth from Beijing in a Mercedes, my abiding impression of Liu Min Ying is that it is a work-a-day place.

Bian Yousheng argues that a key impact of ecological agricultural construction in Liu Min Ying has been the education of groups of ordinary villagers across the country who are aware of environmental issues associated with agricultural production. He cites Zhang Kuichang as a prime example, a man who had been a (relatively) ill-educated young peasant in the 1970s who rose to become an expert on ecological agriculture and who is now able to lecture at institutions up and down the country on the subject and propagandise for it. In the mid-1990s, the name of Zhang Kuichang is well-known and well respected in other eco-villages.

Close on a decade after it won the UNEP Global 500 award, Liu Min Ying continues to undergo considerable development: in June, 1992, on my first visit, the first village level biogas digester had not been built, by 1997, a second digester was within a month of completion. In the period from 1992-6, the central area of the village was completely redeveloped, a 1000 seater auditorium, a new education building and a new civic centre with a guesthouse had been constructed and opened, there were new factories, new vegetable plots, and a new park. There were new tractors and new irrigation equipment. It appeared a well-kept and well-planned village, with designated industrial, agricultural and residential areas.

Liu Min Ying has benefited from a number of advantages, the most obvious being its adoption by Professor Bian Yousheng as his Institute's experimental ecological agricultural site. Though little money came directly from that source, substantial technical assistance was poured into Liu Min Ying from 1982-5, and it has continued to benefit from the watchful eye of Bian Yousheng and from any necessary technical assistance since. Bian You Sheng is a member of the village leadership, is a frequent visitor to Liu Min Ying (I travelled with him there on three occasions) and his face is well-known and popular on its streets, despite living in his work unit in Beijing. It is clear that Beijing (and indeed the National Environmental Protection Agency) has a stake in its continuing status as a model eco-village. Beijing's grant of 1m. yuan in 1996 for the building of the second biogas digester is evidence of that.

However Bian Yousheng had asked many other villages and work-units of their interest in being an experimental site before he came across Liu Min Ying in 1982. The receptiveness of the village leaders was thus also a crucial factor. Indeed, as with many eco-villages, Liu Min Ying seems to have been blessed with particularly imaginative and forceful leaders, themselves blessed with considerable longevity. Zhang Zhanlin, a Mao Zedong look-a-like, became *shuji* in 1970 and has remained in that position ever since. In the intervening period he has won numerous awards, as a model worker of Beijing, a national model worker, a national advanced Communist Party worker and was Beijing's representative to the 6th, 7th and 8th National Peoples' Congresses. However, while Zhang Zhanlin may have serious *guanxi* with the Beijing authorities, it is clear from practical observation that, at least with regard to CEA construction, it has been Zhang Kuichang who has done the business. His air of seriousness and authority, his strong body language, his frequent impatience to get back to work (rather than speak to me), his constant presence in the village (he was there on each of the nine occasions I visited the village, Zhang Zhanlin was there only twice) the evident admiration that Bian Yousheng held for him, all spoke volumes for his role in the initial adoption of ecological agriculture and its continued development.

Both the above leaders and Bian Yousheng *take it for granted* that the successful development of CEA in Liu Min Ying has been predicated on the decision of the village to remain *jiti* rather than divide the land and adopt the HRS. The construction could not have been accomplished in the first instance had every household been allowed to go its own way. CEA is essentially a collaborative exercise given the interdependence of the various elements of the whole. This is particularly true of village-level biogas, where, for efficient operation, inputs (of animal and human nightsoil, stalks etc) need to be collected and outputs, in terms of biogas slurry and the gas itself, distributed on a large, unified scale. Nowadays, unsurprisingly as standards of living improve, individual villagers are increasingly less keen on performing the various operations involved in keeping a household biogas pit efficiently functioning. Moreover, the sort of initial investment necessary in a village-level biogas digester, is most easily raised from a collective accumulation fund. Zhang Kuichang argues that the conditions for successful development of biogas include interest from leaders and the specialist skills of workers, both of which are more likely developed within the *jiti* social form.

When asked about the continuing strength of the *jiti,* Zhang Kui Chang suggests that while it continues to deliver material progress, as it had done so far, it continues to be popular. He is realistic enough to suggest, however, that the younger generation may have different ideas when they take over the leadership in the medium-term future.

The strength of the *jiti,* according to Zhang Kui Chang, is also a function of the strength of the Communist Party branch. As he explained to journalists:

> All the leaders and members of the Communist Party are examples to all villagers. There isn't any unhealthy tendency here, because we have a good party branch. We have been awarded the title 'Beijing Advanced Party Branch' and 'The Capital's Spiritual Civilisation Unit' for ten years in succession... Political and ideological work is the treasure of Liu Min Ying's spiritual civilisation and construction....There are always a lot of meetings in Liu Min Ying....Let us take the meetings of the Communist Party for example: we have party branch meetings once a week, meetings with all the members of the Communist Party once a month, ...discussion meetings every half year, conclusion meetings once a year. We also hold cadres' meetings and masses' meetings regularly, amongst which the meetings of the representatives of villagers twice a year is the most regular one. .. Although meetings are not

115

popular recently in many places, they are still popular in Liu Min Ying, because they are vivid and often solve concrete problems..... No matter how hard the work was, the masses always responded to the party's call. In doing the work of digging channels, planting trees and digging roads, we organised much village labour. Noone complained about it at all..... The consciousness of the *jiti* stems from the high quality of cadres and masses. This quality is the spirit of Liu Min Ying (Zhang,1994).

Whatever the degree of hyperbole involved in the above it is clear that the Communist Party still has an influential role in Liu Min Ying: in 1996, Zhang Kuichang claimed that open village meetings are called twice monthly by the CCP branch and attended by up to 400 people.

Another advantage which Liu Min Ying had at the outset was money. It was a fairly rich brigade by the end of the 1970s which gave it advantages in many respects: it had no particular reason to want to abandon collective organisation, its prosperity reinforced the masses' faith in the leadership, and it could afford, financially, to take risks. Latterly, the success of ecological agricultural construction begat further success as the village's growing reputation on the environmental front attracted financial investment from outside, including abroad. Liu Min Ying's proximity to major municipalities also gave it a head's start: a visit by a dignitary staying in Beijing is a great easier to Liu Min Ying than to an eco-village in the middle of nowhere, markets are closer, technical assistance and institutional support more easily available.

The achievements of Liu Min Ying have their detractors: according to Professor Cheng Xu of Beijing Agricultural University (since 1996 Head of the Scientific Department, Ministry of Agriculture) in May 1993, the economic sacrifices made by Liu Min Ying in developing ecological agriculture have been too great arguing that alternative projects in other villages, such as Doudian, with which he is most closely involved, and others (including Xiao Zhang Zhuang) are better exemplars. My own observations suggest this view is hard on Liu Min Ying: developments in Lu Min Ying are there for visitors to see: the factories described above have been built, two-storey homes have been constructed and the villagers enjoy consumer durables in their homes and a range of public facilities outside them at least to the levels enjoyed in other eco-villages and above them in most. And even if there may be some eco-villages which have made more economic progress, as Cheng Xu suggests, the villagers of Liu Min Ying are materially *far* more comfortable than they were a decade ago.

Liu Min Ying also has its detractors on the ecological front: chemicals fertilisers and pesticides have not been totally abandoned; meanwhile the animals are not all allowed to roam freely, specifically the pigs are in stalls and the chickens kept in mass-production cages. According to Karin Janz, a rural development consultant who spent 1994/5 in CAID in Beijing Agricultural University, and who visited Liu Min Ying last in 1992, the agriculture in Liu Min Ying 'is not the same as the IFOAM organic certified agriculture.'

Despite any detractors, Li Xueming, the 43 year old director of planning and propaganda and hence responsible for organising village meetings, argued in 1993 that there was a general consensus amongst villages on the main aspects of development, accepting that this had not always been the case and that the arguments for biogas and for organic rather than chemical fertilisers had been won amongst the villagers only as a result of the economic benefits

116

perceived to have resulted from them. The commune as a form of social organisation, he argued, is also popular: it is the very richness of Liu Min Ying, in terms of material wealth, which makes people believe in the collective ownership. The gap between rich and poor in the village is also very narrow, he suggests. When in 1993 Li Xueming and Li Caiyun, public relations officer, were asked whether, as industrialisation and services developed further, ecological agriculture would become a thing of the past, they emphatically denied this, suggesting that while fewer villagers would be involved, CEA would remain an important element of the village's economy.

From the evidence of these and other leaders, it seems Liu Min Ying will remain a force for environmentally friendly economic development, based upon good social relations and common ownership for some years to come. Liu Min Ying *jiti* appears in good health in the 1990s. The recent shifts towards a service economy, in particular the development of eco-tourism in the second half of the 1990s, suggests that it is keeping well ahead of the game.

8(2) Xiao Zhang Zhuang

8(2) 1. Introduction

Xiao Zhang Zhuang is a largish village of some 3502 people from 802 households and is one of 28 villages in the township of Xie Qiao in the Fuyang district of Yingshang County in the north of Anhui province. It has a total area of 422 hectares, 324 hectares of which is cultivated land, almost 1.6 *mu* per head, the rest being residential land, industrial land, forest and fishponds. It is situated in the southern part of the north Huai river plain *(Huai Bei)*, an area of China notoriously prone to serious, at times disastrous floods. In the summer of 1991 a well publicised flood led to substantial loss of life and damage to property, while another serious flood in the summer of 1996, though not the cause of death led to substantial property damage (and difficulties in travelling to Xiao Zhang Zhuang). Anhui is traversed by the Huai River *(Huai He)* in the north and the Yangtse River *(Chang Jiang)* in the south and, with a rural per capita GDP of 1808 yuan in 1997 (Statistical Yearbook of China 1998, p.347), is in the poorest third of provinces in China.

Meanwhile, Xiao Zhang Zhuang is in one of the poorest regions of Anhui. It is situated about 50 kilometres from Fuyang, the capital of the district, and 20 kilometres from Yingshang, the county town. The new Beijing-Hong Kong railway opened in June 1997 stops in Fuyang, but the village remains in an isolated, rural location; there are still many parts of Anhui province out-of-bounds to foreigners without special permits and Xiao Zhang Zhuang was, until very recently, one of these. The car journeys from Bengbu, Northern Anhui's largest town and from Hefei, Anhui's capital, both take about 5 hours on uncomfortable roads. Northern Anhui is not only a region of material poverty but also of evident pollution: the *Huai He* itself is surrounded by industries discharging their waste into it: on the road from Bengbu to Yingshang, which crosses the *Huai He* lie any number of large industrial works, including chemical factories and iron works, emitting all sorts of pollutants of various colours and smells into the atmosphere.

8(2) 2. Recent History

Xiao Zhang Zhuang as a political or administrative entity is of very recent origin, being created in 1958 from 11 previously natural, though scattered, disorderly settlements as one of the many brigades *(dadui)* within the Gong Ji commune. From the outset it was very poor occupying at the start large tracts of weed-ridden waste land. During the period of the Great Leap Forward there were frequent floods and droughts, indeed between 1958 and 1960 there were no harvests of any kind forcing large numbers of peasants into a life of begging. The popular saying at the time was 'with rain, floods, without rain, drought.' Moreover, the poor environmental conditions of the time, in large part responsible for the poverty, actually got worse as trees that

118

did exist were chopped down for firewood. The 1960s saw no substantial improvements and productivity and incomes remained very low: grain output was no more than 90 *jin* per *mu* and per capita annual income fluctuated below 100 yuan.

During the Cultural Revolution, although the brigade escaped the more serious chaos experienced in other areas, emphasis was put on political rather than economic issues and the harvests were exaggerated. The villagers today argue that Xiao Zhang Zhuang had little choice, suggesting that 'it was a social trend' to do so, and that models of development such as the 'learn from Dazhai' movement were unhelpful because they bore very little relevance to local conditions. Xiao Zhang Zhuang's circumstances at the time were no different from other neighbouring brigades. Poverty resulted from the poor quality of the soil and the regularity of flood and drought, in turn resulting primarily from the lack of trees. It is also argued that the brigade's leadership was lacking in direction. Indeed it seems the only positive feature of the poverty experienced by the brigade then was the emergence of leaders determined to tackle it.

In 1968, Zhang Jiashun, at the age of 30 became leader of one of the four residential communities within Xiao Zhang Zhuang. He was from poor peasant stock and his academic education had finished at junior school but he had several years practical experience in the fields behind him and had been a member of the Communist Party since 1957. Zhang Jiashun argues it was the poverty of the brigade in the 1960s which first propelled him into leadership positions.

It was not until the 1970s that the village began its economic reconstruction. There were many problems identified: a large crop land area but with poor soils and lack of fertiliser, an unbalanced agricultural economy based almost entirely on grain monoculture, with few vegetables and negligible animal husbandry, a shortage of fuels leading to a conflict over the use of organic material for fuel, fertiliser and animal feed, serious problems of plant disease and pest control and a lack of trees and woodland providing 'a hard micro-environment for farming and living.' (Anon,1993, p.2) At this time economic conditions were equally hard. In the early 1970s, food was in short supply, wheat crops regularly being damaged by flooding and potatoes were the only vegetable crop. The productivity of the land was very low, average grain output still remaining at no more than 90 *jin* per *mu*. There were few pigs and only two cows in the whole village. Of the 700-odd households 100 were beggar households; as late as 1976 per capita income remained as low as 56.3 yuan. Houses were built of mud with thatched roofs and there was no electricity or indoor water supply. Clearly, at this time, there could be no perceived distinction made between the 'economic' and the 'environmental': it was the poor environment which contributed in large measure to the economic misery experienced and thus any economic recovery was predicated on environmental improvement. In the language of the World Bank (1992, p.2), a classic 'win-win' situation presented itself.

According to villagers, the first step was made in 1971 when Zhang Jiashun sold pigs and with the 100 yuan received bought young trees and planted them. The initial motivation for doing so included the realisation that the lack of trees contributed to the poor quality of the soil, to erosion from strong winds and to flooding. Zhang Jiashun argues that he initially began tree planting because of the country-wide exhortations made to do so by Mao Zedong. The economic benefits of tree-planting were not immediately apparent and people opposed to the decision needed persuasion as to its value; this was especially up to 1973 when the economy was not materially better off than it had been in the 1960s. At roughly this time, at the instigation of Zhang Jiashun, all brigade members - leaders and led, women and men - began communally to dig new ditches. In the first instance, while old, waste ditches were buried, new ones in areas most prone to flooding were cut, making it easier to drain waterlogged fields. Meanwhile, the mud from the ditches was applied to the fields, improving soil quality and a programme was embarked upon to build a web of interconnecting channels not only to alleviate waterlogging but to provide an efficient system of irrigation.

In 1974, economic benefits associated with tree planting and ditch digging became apparent. Zhang Jiashun was credited with success, was secured in his position and in 1975 was elected brigade leader *(daduizhang)* and secretary of the Communist Party *(shuji)* without opposition: given the economic state of the brigade there was no competition to take on the responsibility of leadership. By 1978 the economic situation had turned around, encouraging the consolidation and extension of the water web and further afforestation. Yet these environmental improvements were not in any sense consciously 'environmental', nor were they motivated by an understanding of theory. Rather they were born out of practical necessity and reinforced through practice. As Zhang Jiashun admits, he had never heard of the 'environment' until the 1980s nor of ecological agriculture until 1983. Until then they learned only from practice.

Such was the success of Xiao Zhang Zhuang under the leadership of Zhang Jiashun that in 1979, the latter was chosen as a 'model worker', receiving the award in the Great Hall of the People in Beijing and bringing honour to the brigade as a whole.

8(2) 3. Land Reform

With the proposals for land reform in the early 1980s Xiao Zhang Zhuang was confronted with considerable political pressure from above to follow official policy and divide brigade property amongst individual households. For most brigades in Northern Anhui, who were then, as now, often desperately poor, the decision to do so involved no difficulties. Indeed, the earliest pressure towards family-based farming *began* in Northern Anhui: faced with low productivity and poor brigade leadership, the new opportunities presented by the reforms were greeted with enthusiasm in most villages. This was not the case in Xiao Zhang Zhuang, however, largely because of the economic successes that the brigade had just achieved as a result of its *collective* activities and because of the good relations that existed between brigade members and between leaders and led. Today's villagers agree that relationships within Xiao Zhang Zhuang are, and for many years have been, uncommonly good, with considerable faith in the quality of leadership amongst the people after the economic successes of the mid 1970s.

In any event, there was considerable opposition, particularly amongst the leadership, towards abandoning the collective *(jiti)* in favour of the HRS. Even amongst ordinary villagers, support for the reforms was neither strong nor universal, and without the pressure imposed from above Xiao Zhang Zhuang might well have gone its own way and remained *jiti*. Zhang

Jiashun was clearly torn between two schools: a desire, as a good Party member and senior cadre, to go along with official policy, and as a practising leader of a successful collective, to keep it as such. In the end, after many village-wide meetings, it was decided that the fields should be divided amongst households but that the trees, ditches and irrigation system, mostly constructed in the recent past, should remain collectively owned and managed. Zhang Jiashun states that he made that decision because of the problems of managing the trees and waterways that would have occurred if all brigade property had been divided up, recognising the potentially harmful implications for the economy and environment had the trees and ditches gone the same way as the fields. This appears to have been a popular decision. In the first instance, in 1982, contracts were drawn up with groups of houses, by 1983, they were made with individual households *(bao gan dao hu)*.

According to the villagers today, the division of the fields took place smoothly and it is claimed that relations between households are as good if not better than before since there are none of the petty disputes which existed when the land was communally farmed. That there have been considerable increases in the productivity of the land since its division is not in dispute. However, not everyone in Xiao Zhang Zhuang believes that those increases are predicated on the land reforms per se. Indeed, somewhat controversially and surprisingly given his status as a senior Chinese cadre (he is now deputy leader of the Yingshang County Party) Zhang Jiashun argues that the land reforms may not have been necessary and that, given the solidarity of the brigade around the common good, the importance put on approaching the tasks of the brigade with the spirit of Jiao Yulu and Lei Feng (this was consciously done - the statues of both these important figures in the ideology of modern China are erected in prominent positions in the village) and the quality of the brigade leadership, the same results could have been achieved had the land remained in *jiti* control. Nonetheless, the land division remains popular in the 1990s. Equally the decision to keep the trees and waterways in communal hands is still perceived to have been a good one and, amongst the leaders, to have been absolutely crucial. They argue that, in general terms across China, while land distribution has been economically advantageous (even support for this proposition is not universally strong amongst them), it has been environmentally damaging, 'since people can now do what they like,' including growing trees or not growing trees or cutting down trees, as the case may be.

8(2) 4. Ecological Agriculture

As has been stated earlier, the more self-consciously 'environmentally friendly' developments did not occur until 1983 when the decision to begin ecological agricultural construction was made. This decision, made ultimately by Zhang Jiashun, seems to have sparked no great controversy, largely, it seems, because of absolute faith in the correctness of his leadership based on the village's previous experiences. It also appears to have been made with no particular outside encouragement, merely general exhortations from the national and provincial environmental protection agencies, and without any promises of money or technical expertise. As in other ecological agricultural construction projects, its objectives included the search for a more balanced economy, a more rational use of new, environmentally friendly sources of energy and reduced dependence on chemical fertilisers and pesticides. As Zhang Jiashun argues, the enthusiasm to continue 'environmental' developments amongst the villagers was predicated on their understanding that they would bring 'economic' benefit, as the decisions and developments had done from the mid-1970s onwards: there was, and still is, no suggestion that environmentally friendly development should involve any form of economic sacrifice.

Despite the additional parameters involved in CEA, planting trees, digging ditches and making engineering improvements to the water web, thereby providing enhanced soil quality, better drainage and an improved irrigation system, continued to be prioritised. Altogether 20 protective tree belts, of a total length of 40.8 kilometres and composed of a huge variety of different kinds of trees, deciduous and evergreen, broad leaf and needle leaf were constructed up to 1993. Fifty hectares of orchard were planted, as were 8 hectares of bamboo trees. Total forested area, including the tree belts around the village, ditches and roads, by 1993 comprised 113.3 hectares, or 26.2 per cent of the total area, up from 6.9 per cent in 1980. And with regard to the system of ditches, the developments of the 1970s were refined and enlarged upon. Altogether, 20 separate channels, one 10 kilometres long, were dug, all by hand, and 64 pump wells and 2 electric pump irrigation stations were constructed (Anon. 1993, pp.2-6) Trees and water were the foundation of Xiao Zhang Zhuang's improved ecology.

However after 1983 ecological agriculture involved a new dimension: the development of a virtuous ecological cycle based on the adoption of animal husbandry, biogas and fisheries. Animal husbandry was developed, if on a relatively limited scale, so that by 1993 there were 650 pigs and 700 cattle, raised by individual households. More spectacularly, the village began to raise long-haired rabbits, such that by 1993, their numbers had grown to 15,800. At the same time, 15 fish ponds had been dug or repaired, covering a total area of 12 hectares. Between 1976 and 1993, animal husbandry expanded its contribution of a much expanded total village output from 6.6 per cent to 8.73 per cent, while the contribution of the output of fisheries rose from 0.3 per cent to 0.94 per cent Meanwhile by 1993, 49 new household biogas pits had been constructed. Human and animal night soil in combination with stalks and 'waste' biomass from the fields were to provide inputs to the biogas pits producing gas for cooking and lighting (alongside increased use of solar energy for the same purposes), the waste from the biogas pits to be deposited in the fishponds and used as fish-food, the mud and sediment from the bottom of the fishponds dredged and used as rich, organic fertiliser on the fields raising the fertility of the soil. At least, this was the plan.

Improved and more intensified methods of cultivation, including inter planting, and double cropping, mostly of wheat, harvested in June and rice, harvested in September, were introduced. By 1993 there were altogether 10 different modes of intercropping, including cereals and oil crops and cereals and vegetables and the area of land subjected to double cropping had grown to over 89 per cent. The significance of market influences on rural output is underlined by the fact that the proportion of cultivable land under cash crops grew from less than 5 per cent in the early 1980s to 38.7 per cent in 1993. During the same time, 246.7 hectares of wheat-growing land were converted into paddy fields for planting rice.

8(2) 5. Political Economy in the 1980s and 1990s

8(2) 5.1 Political structures

In terms of political administration, there is, in 1997, a village committee of six people, elected from a panel of candidates nominated by peasants and approved by the leaders of Xie Qiao township, for a renewable period of three years. It is from this body that decisions ultimately emanate, although there are committees of three to five people at the 'community' *(xiaodui)* level who may carry out investigations and who are able to express ideas to the village leadership. In 1997 the six people, only three of whom are Party members, include the 45 year old Wang Xueqi, a member of the Party since 1972, leader *(cunzhang)* for the past 9 years and *shuji* since 1996, two deputy *shuji,* Wu Yuqin, leader of womens affairs and Wang Qing, leader of the local Peoples Army. The village leaders meet 'very regularly' in the newly built village offices, formally at least once a month and there are village-wide meetings once a year. Despite the distribution of the land in the early 1980s, there appears to be an impressive degree of social solidarity and the *dadui* as *danwei* (the responsible work unit) remains influential. There are currently 51 Party members in Xiao Zhang Zhuang.

8(2) 5.2 Economic developments

The productivity of the land has expanded: the total grain in 1993 was 2.97m. kilograms, up from 1.34 m kilograms in 1976, an increase of 220 per cent on a lower acreage. Grain yields per *mu* have increased dramatically, from an average of 140 kgs. per *mu* in 1976 to 400 kgs. per *mu* of wheat and 600-700 kgs. of rice per *mu* in 1993, and this was achieved alongside reductions in the application of chemical fertilisers and pesticides. The rice crop is particularly important economically: residues in the rice stem are used as fertiliser on the fields and as an input, sold outside the village, in the making of paper.

While agricultural output has expanded in volume, satisfying the wants of the villagers, providing a surplus for sale and higher levels of income, the importance of farming in terms of its relative contribution to total village output has declined: in 1976 the proportion of total production accounted for by planting crops was 86.1 per cent, by 1992 this had fallen to 43 per cent, by 1996 to 30 per cent Between the same years, the contribution of forestry to total output rose from 2.2 per cent to 9.0 per cent to 20 per cent, of animal husbandry from 6.3 per cent to 8.73 per cent to 10 per cent and of fisheries from 0.3 per cent to 0.94 per cent to 1 per cent. The increased contribution of forestry is particularly significant. It appears that Xiao Zhang Zhuang still owes a considerable debt to the original decision to grow trees and to keep them communally owned, if not managed. In particular the growing of bamboo has been an enormous economic boon. There are now 18 hectares devoted to bamboo; the land is owned by the *jiti* although the trees are managed privately under contract by 80-90 individual households (10 per cent of the total). According to Shi Lixue of the Fuyang Environmental Protection Unit in 1993, in that bamboo trees grow quickly and have a ready market outside Xiao Zhang

Zhuang at 80 yuan per tree, yet can be used for construction in the village and as raw material in the furniture factory, the bamboo forest represents a 'Green Bank'. This phrase is itself highly a significant one: at no stage has the environmental development of the village been, or seen to have been, at the expense of more immediate 'economic' benefit.

Despite expansion of forestry and animal husbandry in Xiao Zhang Zhuang, the major economic transformation has resulted from industrialization. In 1976 the proportion of sideline production to the total output of the village was, at 1.5 per cent negligible. By 1992, this figure had jumped to 38.3 per cent by 1996 'more than 50 per cent' (although this includes animal husbandry). While in the 1970s everyone worked in the fields, by 1993 60 per cent of the labour force of the village was, at some stage or another during the year, engaged in manufacturing or sideline production, by 1997, 'almost everyone' was. This rapid development took place steadily throughout the 1980s and early 1990s so that by 1997 there were altogether 13 different sideline enterprises. These enterprises include a factory which makes hand-cut and sown quilts in partnership with a Shanghai company that chooses the designs, provides the raw materials and markets the output into international markets. It employs 200 young women on a regular basis and pays them 300-400 yuan a month, depending on production levels. Other enterprises include a furniture making factory, a cannery, a high-grade rice mill, a clothing factory, a melon seed processing factory as well as a brickworks and construction and transport enterprises.

Two new factories, one making paper boxes and the other making plugs opened in 1994 while in 1995, a bottle factory was constructed (on 21 *mu* of previously arable land) using 5m yuan from the village accumulation fund. By 1996, it was already the most important factory in terms of output, profit and employment in Xiao Zhang Zhuang. In 1997 it employed 107 workers, mostly males from the village who earned 400- 500 yuan a month. The factory operates throughout the year and the workers are full time, unlike most workers in other factories which shut down when it is busy in the fields. Its new production line (with its own electricity generator) produces 55,000 bottles a day (for beer and *baijiu)*, with an output value of 9m.yuan, making a profit of 1.1m.yuan and paying taxes of 800,000 yuan to Xie Qiao township in 1996. Working conditions involve extremely high temperatures and overt lack of health and safety regulations, however.

In early 1996, a chemical fertiliser factory was opened employing forty workers and producing fertiliser for units across the entire county. Zhang Zhi-an, *shuji* of Xie Qiao (and son of Zhang Jiashun) explained that the artificial fertiliser, known as *fuzhi suan,* and developed in Langzhou incorporated the latest biotechnology and was considerably better for the soil than normal artificial fertilisers.

All enterprises have been established communally by the *jiti* but some operate through a contract system *(chengbao)* which may involve several families working together *(hehuo)* or a household individually *(jiating).* In the case of the bedding and bamboo factories (and long-haired rabbit factory, shut down in 1995), they are run by three families working together, the first two contracted to pay 100,000 yuan to the village, the latter to pay 500,000 yuan. In addition to the above enterprises, there are a number of single household enterprises principally providing retail and catering services. Some households have got a lot richer than others, have built better houses and some have bought cars. One family has bought a lorry, another a *miandi* (a mini-bus taxi). According to *shuji* Wang Xueqi, despite a few complaints, the contract system works well. He argues uncynically that China's national policy is now to allow some families to get rich first so that everyone else will eventually get rich: he does not see the contract system as a threat to the *jiti,* rather that the *jiti* 'makes things happen' and that individual households remain under its leadership.

While Xiao Zhang Zhuang is keen, in common with other villages in China, to attract foreign capital through the medium of joint ventures it has so far failed to do so. This is unsurprising given its relative inaccessibility. Even as late as 1976, Xiao Zhang Zhuang remained an exceedingly poor village, where the range of economic activity was narrow, productivity of both land and labour low, and where average income per head at 51.3 RMB per annum, earned exclusively on the fields, (i.e. 1 yuan per week!) was so meagre that 15 per cent of the households were forced into a life of begging. By 1997, the situation has been transformed. The villagers of Xiao Zhang Zhuang, with a per capita income of 2,560 yuan in 1996 may still not be rich in comparison to villagers in some other, more prosperous provinces in China but substantially so in relation to other parts of rural Anhui, such as the Fuyang district, where average income per head was only 1100 yuan in 1995. The average for Xie Qiao township in 1995 was 1,650 yuan.

8(2) 5.3 'Social' developments

The most obvious manifestation of improvements in standards of living enjoyed by the villagers has been the development of housing. In 1980, for most villagers, homes meant mud huts without electricity or domestic consumer durables. In the early 1980s, new, single-storey brick-built houses were built with bedroom, living room and kitchen inside, with backyard and toilet at the back and a small patch of garden with trees and shrubs at the front. The standard of housing which these houses represent in comparison to the standards and size of much housing in other parts of China is reasonably high, if scruffy.

However, in 1992, financed in part by the surpluses from its industrial enterprises, Xiao Zhang Zhuang embarked upon a new housing drive with the intention of replacing all the single-storey houses with two-storey houses of substantial proportions, having 3 living rooms downstairs, 3 upstairs, toilets upstairs and down, all with balconies. By June 1993, 37 new homes had been built at a total cost of 1.2 m. yuan The original intention was to replace the single-storey houses with 2-storey houses, at an average cost of 40,000 yuan per house, by the end of the decade. All these homes, unlike the single-storey homes, were designed to use biogas, with high efficiency stoves and solar-energy panels. However, though square footage is very generous and despite extended access to running tap water from underground wells, the provision of utilities inside is not significantly in advance of the single-storey houses they replaced. Despite this, the new housing stock is the showpiece of the village and the most immediate and obvious illustration of its economic and social development. The leadership take visitors to see the new homes as a priority; meanwhile there is almost complete unanimity amongst the villagers that improvements in housing represent the most important 'social' benefit. There are now 30m^2 of housing per person in Xiao Zhang Zhuang. Between 1993 and 1997, however, there have been no major additions to the new housing stock.

Higher levels of welfare are manifested in much higher levels of consumption within the household. Diets are now considerably diversified, as more meat, fruits and vegetables are consumed while households routinely include a range of consumer durables such as colour televisions and electric fans, and in some cases, video-recorders and refrigerators as well as such items as bicycles, clocks and cameras which would have rare even as late as 1982.

Other social benefits include significantly expanded educational opportunities. In 1970, few children had the benefit of more than 2 years primary education.In the late 1980s a three-storey primary school, with 5,400 m^2 of floor space was constructed now attended by all children of primary age. In 1997, a middle school was opened. Children at the age of 12 now routinely attend county middle school, some with realistic hopes of continuing their education after finishing their studies there. There were, in 1995, 7 children attending institutions of higher education with the support of the village (few, if any, of whom will return to the village after graduation).

In terms of provision for old people, it was impressed upon me how long life expectancy was in Xiao Zhang Zhuang, up to the national average of 68. In 1995, there were 20 octogenarians, and some were in their 90s. As is still common in China, children, particularly male children, are expected to look after their aged parents, but the 27 old people in the village without children are housed in one of the two old peoples homes rebuilt since 1980. There are four clinics in the village which provide 'inexpensive' medical care and if villagers are seriously ill and need hospital treatment outside the village, they will be subsidised from village funds. Su Congfu, one of Xie Qiao's leaders, insisted that (by implication, unlike Britain, which he had just visited) the 'Chinese tradition was to respect old people'.

Meanwhile throughout the last ten years or so, the village leaders have attempted to apply China's strict birth control regulations: as a result the natural rate of increase of the population of the village has fallen from 10 per thousand in 1986 to 8.6 per thousand in 1995. The adherence to the one-child policy was 86 per cent in 1995. Wang Yuqin, leader of women's affairs is responsible for enforcement: married women of child-bearing age are regularly tested for pregnancy and anyone with a second conception is pressurized to have an abortion. In some cases, a woman may be allowed a second child, but thereafter she will be forced to undergo sterilization. The leaders admit that the enforcement of the policy does cause some difficulties.

In the late 1980s and early 1990s, in order to expand opportunities for recreation and leisure, Xiao Zhang Zhuang built a pleasure park. This involved constructing an artificial hill 50 metres high on top of which is set a traditionally constructed and decorated gazebo approached by a series of steps and providing a view of the intensively tree-covered surrounding area. The total area of the park covers nine hectares and includes within it a plant nursery, a recreation ground and a small zoo, the latter housing, in 1993, a number of rather sad-looking animals. The plant nursery grew over a hundred types of valuable trees and flowers, including cedar, cypress, sweet osmanthus, peony, azalea, plum blossom and

126

narcissus and had, in the previous 5 years earned an average income of 18,000 RMB per year (Anon, 1993, p.3).

In 1995, the village began upgrading of the pleasure park to include a childrens playground area, a much extended zoo with a unit for birds, a huge pagoda and a boating lake with a marble boat aka Summer Palace. There is also a large sculpture, completed in 1997, with calligraphy extolling environment protection from, amongst other, Jiang Zemin, Qu Geping and Mostafa Tomba. The money for this project (1m. yuan) was provided by a Hong Kong businessman, Hu Wenbin, who had had business dealings with the village (he had bought rabbit hair from the factory in Xiao Zhang Zhuang), was against pollution and interested in ecological agriculture. The new park opened in 1996 and by 1997 was attracting visitors from far afield who paid 10 yuan each for the privilege of enjoying it.

8(2) 5.3.1 Some personal accounts

(i) Interviewed in 1993, *Han Weixi* was a 43 year old male peasant, married with two children, boys aged 13 and 10, the older in middle school and the younger in junior school. As a boy Han Weixi had only 6 years of formal education, 4 years at primary school returning to study for two years at the junior middle school at the age of 19, graduating from the latter in 1970. He lived with his family in a seemingly comfortable single storey brick built home with three rooms, a kitchen, a back and front yard in a pretty, silvan location. He was one of the minority of his age in the village who worked only in the fields. His household was allocated a total of seven *mu* in the land reform which he still tilled, half of which was given over to rice and half to vegetables. There was no intercropping on his land. He said that in 1970 conditions in the village were very bad. He lived in a mud hut without electricity at a time when household consumption of grain was at most 200 kilograms per year (little more than pound of rice per day for the entire household), 'not enough'. Enthusiasm for the changes wrought in the 1970s amongst the peasantry was warm and he suggested that land distribution was enthusiastically adopted in 1982. Asked what he had thought about land reform at the time, he replied he 'didn't think', simply obeyed the leadership. Nonetheless, along with the general view, he was very satisfied with the results. Before the land was finally distributed there had been lots of negotiations, bad land was put together with good land and most peasants felt the final distribution was fair. He felt that social solidarity was strong after land distribution and remembered there was considerable mutual aid and the sharing of tools and property. He suggested that currently there was still overwhelming support for the land reforms, land yields per *mu* being so much greater than they were. He and his family were able to grow the food requirements of his household and he sold his surplus grain in the open market, either at the grain station in Xie Qiao or locally. He recalled that it was after 1984, after electricity became available, that he began to buy a television and other consumer items. Since then, his standard of living had increased year by year.

Asked about the changes in the environment, it was clear the word had no particular mystique for Han Weixi: he was, however, aware that the village looked more beautiful now with trees and flowers and he was very proud of this. He did not comprehend the notion that economic development and environmental protection could ever be somehow in conflict. Asked what he considered the most important change to have been, he replied housing and was, despite the apparent comfort of his present home, looking forward to moving to a new two storey home soon. His parents and his wife's parents were all still alive. The former were living with his younger brother but planned to move in with his family when the new house was built. Han Weixi's extended family, numbering 16 members, all lived in Xiao Zhang Zhuang. His family had not used biogas but planned do so when he moved to the new house. Asked what he considered to be the main reason for the improvements that had taken place, he replied, (after some prompting from our 'minder' yet with apparent conviction) strong leadership.

Interviewed again in 1995, Han Weixi was still in his one storey home although his new home was apparently nearing completion, being built by the *jiti* at a cost of 40,000 yuan, half of which he had to provide from his own savings. He felt things were still improving: he farmed eight *mu* of land now, in two plots, and had bought a small machine for sowing. He had sheep, geese, chickens, ducks and rabbits and used the night soil as organic manure on the fields: he still used chemical fertiliser, however, despite its increasingly high cost, and pesticides, particularly on the rice crop. He had introduced double cropping, wheat and rice, to his land which now yielded 600 *jin* per *mu* of wheat and 1400 *jin* per *mu* of rice. He could sell the rice in the open market at 1.30 yuan per *jin*. Still one of the few villagers to farm full-time, he suggested it was possible that his plot would become bigger in time as villagers gave up the land. He was helped in the fields by his wife and children when the latter were not at school.

Interviewed again in 1997, Han Weixi and his family were *still* in their single-storey home, although they planned to move next year. He was no longer a full-time farmer and worked, alongside his wife in the new pleasure park, he as a labourer, she in charge of the car park, their schedules depending upon the work in the fields. Han's now 17 year old son had graduated from middle school and was self-employed selling pigs. He bought the pigs ready for sale from local households, hired a truck and transported them to Anqing (Anhui's port on the Yangtse). He viewed this work as temporary and expected to get a job in an enterprise before too long. Han's younger, 14 year old son was still at middle school, and hoped to move on to senior middle school in 1998 and to college when he turned 18.

Han Weixi's land, including private land *(sirentien)* and contract land *(kouliangtien)*, remains at eight *mu.*, planted with rice, wheat and vegetables. His household's contract to the state is 100 jin of rice per *mu* but since his yields are above 500 kgs. per *mu* from two harvests, he has plenty left for his family's consumption and for sale in local markets. He keeps two pigs, three sheep and 30 chickens, from which he derives fertilizer for his fields. He still uses chemical fertilisers, however, for up to 30 per cent of his needs, and chemical pesticides. He and his wife's incomes are more than enough to cover his living expenses and he estimates they can save up to 12,000 yuan a year. The family seems pretty content.

(ii) Interviewed in 1993 *Gao Yurong* was a 45 year old woman, the wife of Han Weili, it was perhaps instructive that when I asked her her name, through my interpreter, she introduced herself by the name of her husband, still, apparently, a common tradition in rural

Anhui. She had two children, a girl aged 20 and a boy aged 18. Her daughter was studying at the county medical school with the intention of becoming a nurse, her son had just graduated from junior school and was hoping to go to technical school. She lived with her family in a comfortable single-storey three roomed house with a small courtyard, only a short distance from the home of Han Weixi, which was built for her in 1982. She worked with her husband in the fields: they had six *mu* of land planted with rice, wheat and vegetables. She remembered life in the 1950s and 60s as being 'very poor', there being frequently 'no rice'. During the period of the Great Leap Forward, from 1958-60, she ate in the communal dining hall which kept people from starving. In 1962 there was a form of land distribution which was not successful and only lasted a year. She was at junior school at the start of the Cultural Revolution in 1966, graduating from junior school in 1968 at the of 19, having started formal schooling at the age of 11. As an eighteen year old school student in 1968, she took part enthusiastically in the political movements of the Cultural Revolution. In 1970, she remembered that living conditions were still very poor and that when she got married to Han Weili in 1973 they lived, in common with the other villagers, in a mud hut.

She suggested that living conditions gradually began to improve only in the mid 1970s as a result of 'strong leadership' and 'the rules of the state'. Everyone was enthusiastic about the changes and took part in digging ditches. Productivity had already risen substantially because of the 'good cooperative economy' engendered by Zhang Jiashun before land distribution in 1982 which was also enthusiastically embraced. 1993 yields of 500 kgs. per *mu* of rice and 350 kgs per *mu* of wheat allowed her and her husband to sell their surpluses of grain and vegetables in the free market. Since the extension of electricity to her home in 1985, she had been able to enjoy an increasing range of consumer durables. She agreed that improved housing had been the single most significant development in Xiao Zhang Zhuang in recent years and although she was looking forward to her new two-storey home which she planned to move to in about two years time, she was happy where she was and would be sad to see her present home knocked down.

Asked about environmental protection, Gao Yurong replied that it was very important to her. She was able to reflect on the irregularity of floods and droughts in the last ten years and mentioned that in 1991, in the year of disastrous flooding in Anhui Province the situation had not been serious in Xiao Zhang Zhuang because of the superior drainage system. Though she had no biogas or solar panels in her home presently, she would do so in her new house. (Perhaps it will be just as well: the interview was cut short because of a power cut in her home).

Interviewed again in 1997, Gao Yurong still lived, with her husband, son and parents-in-law in the same one-storey home and was having it extended, rather than move to a new one. Her older daughter was now 24, married and working with the handicapped in Xia Qiao hospital, her son living at home and working in the grain station in Xie Qiao township She still worked seven *mu* of fields with her husband but additionally earned money raising and selling pigs. Their household contract was to supply 100 jin of rice per person per *mu* to the state and since their seven *mu* yielded approx. 7000 kilograms of rice and wheat a year, they had plenty over for sale in local markets. Gao Yurong had a further 'sideline' activity, one that she had practised for 28 years, ever since qualifying as a doctor: she ran a clinic, dealing with minor medical problems using western medicine. She had been a barefoot doctor during the Cultural Revolution. With 5-6000 yuan coming from grain, their total income was in the region of 10,000 yuan a year, and since their parents-in-law earned approx. 10,000 yuan (her father-in-law is deputy *shuji*), she felt very well-off.

(c) Interviewed in 1995, *Mian Yang* was a 39 year old woman born in Sichuan, graduated from junior middle school but had met her future husband through a go-between and had come to Xiao Zhang Zhuang in 1979 to get married to him. The first child, a boy, was born in the same year. She remarked that even in 1979, living conditions were much better in Xiao Zhang Zhuang than in Sichuan and that they had much improved since then. Her 16 year old son attended middle school and her 12 year daughter primary school. Her husband was a village leader and was currently in charge of the pleasure park. She had a contract to work 20 *mu* of land, 14 *mu* of which were devoted to growing fruit trees, primarily peaches and grapes, the only grapes in the village. During the picking season, the whole family including her childrens' friends got involved, but she was responsible for running the business and fulfilling the contract, in her case to pay a tax of 200 yuan per mu annually, and she started work at 4am every morning to ensure she did so. She estimated she earned 20,000 yuan a year selling fruits in local markets. The other 6 *mu* were planted with crops and vegetables mostly for her own family's use. She owned 3 pigs, some chickens and ducks and used the night soil of the pigs as fertiliser, supplemented by a small amount of chemical fertiliser, sprayed the fruit trees and crops with pesticides and made use of the irrigation provided by the *jiti*. She lived in a new home, built six years ago for 15,000 yuan, burned coal in a stove for cooking and used electricity for lighting. For the coal she paid 30 yuan a month, for the electricity, 10.2 yuan a month. Other expenses paid to the *jiti* included 170 yuan a year to pay for her son's education, 70 yuan for her daughter's. This still allowed her family to grow rapidly richer. It was impressed upon me by my hosts that Mian Yang was a very capable woman.

8(2) 5.4 Environmental Developments

Trees have provided the foundation for any environmental benefits enjoyed by Xiao Zhang Zhuang over the last 25 years. The barrenness of the environment and the low state of economic development in the late 1960s provided a 'win-win' opportunity as the shortage of biomass for cooking led to the felling of trees and to the degradation of the environment that deforestation implied in an already denuded landscape.

Forest cover in 1997 was 26.2 per cent of the total area of Xiao Zhang Zhuang, the trees growing in designated wooded areas, mostly on the surrounds of the village, alongside roads, pathways and ditches, in protective belts between fields, in 'community' focused spaces close to homes and in parks. Trees of all kinds exist, coniferous and deciduous, bamboo trees and fruit trees. In 1993, the village owned 167,000 trees, 51 for every man, woman and child. And the benefits of this tree cover in immediately environmental terms are varied. Trees have improved the 'cropland microclimate and raised the ability of the crop land to resist natural calamities': the protective belts of trees have reduced average wind speeds (it is claimed) by up to 47 per cent, significantly reducing soil erosion, they have reduced rates of evaporation by 18.9 per cent, increased the air humidity by 7.1 per cent and the temperature of the soil by 1 °C. According to the local Fuyang Urban/Rural Construction and Environmental Protection Department, these effects are alone responsible for increases in the grain yield of between 8 and 12 per cent. Meanwhile, the trees have provided a 'good perching environment for birds' and as a result a great variety of birds have been attracted there. According to the investigations of the Fuyang Department, in 1993 there were 33 different kinds of birds from 20 different families. And the increase in bird life has not merely led to a more pleasant natural environment but has to some extent inhibited the activities of natural pests, thus reducing the need to apply pesticides. Meanwhile in 1993 the trees dropped 1.34 m. kgs of leaves which were used for animal feed and organic fertiliser (Anon, 1993, pp.1-12).

The trees have, in addition, more straightforward environmental advantages: they provide shade and beauty. At a local level, this has meant, for Xiao Zhang Zhuang, a more pleasing natural environment, appreciated by villagers, something which is not that common in the Chinese countryside in general and downright rare in Northern Anhui province in particular.

Soil quality has improved as a result of additional tree cover. The application of organic fertilisers from the night-soil of animals and, temporarily, from biogas slurry, has helped too, reducing the application of chemical fertiliser from the levels of the early 1980s. Soil fertility has demonstrably improved: the organic matter content has risen from less than 1 per cent in 1983 to 1.63 per cent in 1993 and the alkaline nature of the soil has been neutralized, with the ph factor of the soil falling from 7.4 to 6.7.

Other developments, in particular the increasingly sophisticated water web, composed of interconnected trenches and dykes, involving pump wells and electric pump irrigation stations, has significantly improved the systems of drainage and irrigation, ensuring good harvests despite flooding or drought.

In the early 1990s, 49 biogas pits were built but despite the high profile of biogas planned in the ecological development of the village they were built over in 1994. All homes have been lit by electricity since the early 1980s and cooking is performed by burning stalks or coals in new 'energy saving' stoves, with some boiling of water using solar energy. Shi Lixue of the Fuyang Environmental Protection Department confirmed that electricity remains a popular source of energy because it is presently so cheap and that biogas is unlikely to be a major energy source in Xiao Zhang Zhuang until the price of electricity, currently administered by the authorities, and not reflecting its true costs, is higher than it currently is.

The substantial achievements of Xiao Zhang Zhuang in the environmental and economic fields were formally recognised by the decision of the UNEP to grant the village the Global 500 award in 1991. These awards are based on the recommendations of China's National Environmental Protection Agency in Beijing and while the latter became increasingly impressed by the achievements of Xiao Zhang Zhuang in the 1980s the village was not formally recognised for its achievements outside of the county until 1989 when it was nominated as a 'civilised village' across China and got favourable coverage in the press, radio and television. Additionally, Zhang Jiashun was, in 1989, awarded for the second time the all-China distinction of 'model worker'. As he had done ten years earlier, he received the honour in the Great Hall of the People. Further honours were heaped on Xiao Zhang Zhuang when, after investigations by the Anhui Environmental Protection Agency and NEPA, the latter invited Mostafa Tomba, executive director of UNEP to visit Xiao Zhang Zhuang in September 1990. Subsequent to the visit, Tomba wrote that the village provided a 'good example for the whole world in the protection, improvement and development of the eco-environment and it has found a model of development for those countries with underdeveloped agriculture' (the village visitor's book, 1993). In 1991, the UNEP included Xiao Zhang Zhuang amongst its 'Global 500 roll of honour' for environmental achievement with the following citation:

> for 20 years the people of Xiao Zhang Zhuang have been practising sustainable agriculture. By developing irrigation and drainage systems and planting trees to reduce soil erosion, they have succeeded in raising the soil fertility. They have initiated an aquaculture industry, are utilizing marsh gas instead of domestic fuel and have reduced population pressure on the environment by successful family planning. Through sustainable development, this village will ensure a future for their rural economy and future generations (UNEP, 1993).

The announcement of the award was accompanied by local celebrations and a report by Anhui province to all people to 'learn from Xiao Zhang Zhuang'. Villagers profess that they felt great pride in the achievement. Zhang Jiashun received the award on behalf of the villagers of Xiao Zhang Zhuang in Stockholm, Sweden,in June 1991.

In the following year, in 1992, Zhang Jiashun was, as an individual, himself the recipient of a further Global 500 award. (This is unprecedented: China has won only 17 of these awards in total since 1987. For one enterprise, in effect, to win two is remarkable). According to the UNEP,

> Mr. Jiang Jiashun led the villagers of Xiao Zhang Zhuang in Yingshang County, China, in the mid-70s in eco-construction through self-reliance without any state funding. This is a model village based on comprehensive eco-development principles for integrated development of agriculture (crops and animal husbandry), forestry, horticulture, fisheries, water resource management, energy production, social infrastructure etc. which has been accomplished with resources mobilised entirely by the people of the village, with technical support from the government agencies. In 1978 when the villagers launched the reconstruction programme the village was a poor wasteland, frequently subjected to floods and droughts, and dominated by a low productivity monoculture, barren vegetal cover, lack of fuel, replete with plant disease and insect pests aggravated by use of chemical pesticides. The main measures effected by the villagers in the reconstruction programme were (a) soil improvement and harnessing of water for irrigation,(b) village planning (including improving domestic water quality), (c) biological reconstruction through tree planting and afforestation (d) animal husbandry and agricultural production structure was transformed with the introduction of biogas production, together with firewood saving stoves (UNEP 1993).

Zhang Jiashun was presented with the award by the President of Brazil during the UNEP Earth Summit in Rio de Janiero on June 5, World Environment Day, 1992 and has subsequently been included in International Who's Who.

8(2) 6. Conclusions

In the last twenty years or so, Xiao Zhang Zhuang has made observable and impressive economic advances at the same time as, perhaps *because of* paying attention to improving the quality of the natural environment. It is a well-known, high profile village within CEA circles and it remains a model of rural development in Anhui. Nonetheless, it is not one of the richer eco-villages, the housing drive of the early 1990s came to an end in 1993 and in the mid-1990s there appear to be certain question-marks over its commitment to environmental protection in general and to CEA in particular.

Over the years, the village has benefited from strong, capable and respected leadership. Zhang Jiashun, though not formally a member of the leadership of the village since 1979, presently being deputy chief of Yingshang County, lives in a residence in Xie Qiao, a few minutes walk from Xiao Zhang Zhuang and clearly still rules the roost there. His name is on everyone's lips; when villagers refer to the strong leadership of the village they are referring to Zhang Jiashun. Even a teacher in the middle school referred to him in awesome tones as 'our great leader' (speaking in English, which most of the assembled company could not understand). This is clearly partly the result of respect he has earned from making and executing progressive decisions in the past. But it is also due to his power as a high ranking cadre, his local fame (I gained reflected glory when he came to the bus station in Yingshang to see me off) and his occasionally fiery temper.

Another advantage of Xiao Zhang Zhuang has been a strong sense of social solidarity, the result, perhaps, of the communal successes earlier achieved: indeed, by the early 1980s it was one of the richest, most economically developed brigade in the region. In many other respects, however, it has been badly off.

It was never wealthy and did not have a rich seam of accumulated savings to fund ecological agricultural construction, nor has it benefited from outside assistance. It is situated in the back of beyond, in a remote, poverty stricken part of rural China a long way from any centres of populations or rich markets. It is still not well endowed with transport links (although the construction of the Beijing-Hong Kong railway line through Fuyang in 1997 may reduce its isolation). And while it kept its water-web and trees in collective ownership after the rural reforms, the *jiti* was weakened with the division of the fields amongst individual households and the introduction of the HRS.

It is arguable that the above disadvantages have combined to put question marks over the continued successful operation of CEA in Xiao Zhang Zhuang. Bian Yousheng suggests that Xiao Zhang Zhuang was a questionable winner of the UNEP Global 500 award in the first place and explains that the village is not ranked any more in national terms as a model ecological agricultural system primarily on the grounds that unacceptably high levels of chemical fertilisers are applied. By observation in the village from 1993-1997, it does appear that firm commitment to the principles of CEA, particularly toward the construction of an archetypal CEA system, and despite protestations to the contrary from village leaders, is on the wane.

Perhaps the most obvious manifestation of this is the abandonment of any commitment towards the generation of biogas. The 49 household- level biogas pits, newly built and proudly shown to me on my first visit in 1993, had by 1994, been built over, leading both to the abandonment of biogas production, necessitating continued burning of coal and stalks for cooking as well as increased reliance on electricity in the home and to the reduced availability of organic fertiliser.

Successful biogas operation, above all, requires collective organisation. Individual farmers throughout China have largely abandoned household level biogas pits and the building over of those that existed in Xiao Zhang Zhuang in 1993 appears unregretted. Biogas production at the community level requires a very well organised system of collection of night soil, a high level of commitment at the technical level to ensure efficient output and, if this is to be done using a community level biogas digester, a substantial initial financial outlay. All of these conditions are more likely to result from a strong *jiti* with control over animal husbandry and planting. Yet in Xiao Zhang Zhuang, animal husbandry which developed more slowly than in other eco-villages and was never managed by the *jiti*, all farm animals are owned by individual households and the resulting excrement is collected by those households and used directly on individually farmed plots of land. In 1995 Zhang Jiashun recognised that biogas reconstruction would involve a more substantial role for the *jiti,* particularly in the construction, maintenance and repair of the new pits he envisaged but he did not have a community level biogas digester firmly in his sights. In 1996, Xiao Zhang Zhuang was no further down the line in biogas development: indeed he seemed unapologetic, arguing that biogas *'only'* had three advantages (the saving of electricity, its use for cooking and as a fertiliser) and that its further development had become less vital with the raising of living standards and the greater availability of electricity. He did envisage a communal biogas digester in the fullness of time and was aware that since it involved science and technology it would need to be constructed and run by the *jiti,* but in 1996 there were no definitive plans in the offing. In 1997, Su Congfu, Zhang Jiashun's right-hand man, was actually *dismissive* of biogas, explaining that it was *'hen duo mafan'* (lots of trouble). Biogas is clearly off Xiao Zhang Zhuang's agenda.

Other illustrations of the village's waning enthusiasm for CEA are evident. The proportion of chemical to organic fertiliser used on the fields is disputed (Bian Yousheng claims 30 per cent and rising, village leaders claim 10 per cent).There are no obvious plans to grow either 'green' or 'organic' vegetables. Meanwhile, the policy of disallowing polluting industries is a policy of Xie Qiao township, which is enforced on Xiao Zhang Zhuang. And according to Zhang Jiashun's son, Zhang Zhihui, this policy *had* to be enforced by the township when the leaders of Xiao Zhang Zhuang wanted to set up a paper-making factory in 1994. The building of a chemical fertiliser factory in 1996 does not bode fair for the future consolidation of CEA. And the obvious (if understandable) enthusiasm for the motor car (houses are already being built with garages in anticipation) suggests some lack of environmental awareness.

Whether Bian Yousheng's remarks about Xiao Zhang Zhuang are entirely accurate, there are clearly certain difficulties facing the village and it is not obvious that CEA will continue to provide the basis for further economic development, despite its power as a propaganda tool and its use as such. On the positive side, its reputation to date has attracted money from Hong Kong to build the new pleasure park and hence to develop tourism, (although the leaders themselves never use the term eco-tourism). However Xiao Zhang Zhuang appears to have reached the position in which there is some conflict perceived between further economic progress, narrowly conceived in terms of material welfare, and environmental improvement, *while income per head is still relatively low,* and there are difficulties facing Xiao Zhang Zhuang in making advances on both fronts.

As has been noted, Xiao Zhang Zhuang is situated in an area of relative poverty making, in common with many other parts of rural China, very rapid economic development frequently without any concern for the quality of the natural environment. It does not benefit from being well linked to cities with large populations or rich markets. It has no connections with scientific or technological institutes from which it can gain further money, technical help or inspiration. And its social organisation, with a *jiti* stronger than most, yet with a considerable element of contracting-out of production, both agricultural and industrial, to individuals, households or groups of households, leaves itself exposed. On the one hand, some families *are* demonstrably getting richer than others, fuelling some discontent with collective norms, yet on the other, the *jiti* is not positioned to advance a model ecological agricultural system.

Xiao Zhang Zhuang significantly expanded agricultural yields in the 1980s partly because of the *jiti* owned and managed water-web and irrigation system, and better cropping systems coupled with expanded fertiliser use, both chemical and organic, as animal husbandry was encouraged. But the latter was never developed by the *jiti* itself, so there was no significant expansion while the small size of private plots is inimical to much further expansion of yields as large machines are ruled out and economies of scale unforthcoming. Already, there is drift from the land as factory work promises higher incomes: Han Weixi's family now earn more from enterprises and sidelines than from the land (and he was full-time farmer as late as 1995). In these circumstances, it is hardly surprising that the CEA is taking a back seat.

The situation is perhaps made worse by the fact that in other neighbouring villages, some families are becoming seriously rich. In Xie Qiao village, for example, a few kilometres from Xiao Zhang Zhuang, the *jiti* was abandoned in the early 1980's and almost all production (80/90 per cent) is in 'private' hands. Private business has become very important (it was claimed that people have been doing business and market trading there for 100 years) and the village has recently become renowned as a centre for the production of long-haired rabbit fur. In recent years, the rabbit fur business has particularly well, with prices at 280,000 yuan per ton, bringing in altogether 70 households into the rabbit hair business in some form or another in Xie Qiao village alone. For example, Xie Chaoyi (whom I visited in 1995, 6 and 7) started in business in 1982, selling rabbit hair initially to Guangdong province and now exports to Korea, Japan and West Europe. He presently owned 4 factories employing 58 workers, and made, after taxes and all costs, 2m.yuan profit in 1996. With accumulated savings of 10m yuan, he claims his family is one of the richest in Xie Qiao and is known as a local *daquan* or *dahu* (moneybags). Zhi Daoqin is another factory owner, in the rabbit fur business for the last 6 years who made 200,000 yuan in 1994. Both men have built new, large two-storey, quadrangle houses and own large black cars.

Already, some families - for example, those responsible for the rabbit-hair factory - are emulating their success in Xiao Zhang Zhuang. Xie Chaoyi argues that his wealth is not the source of jealousy amongst other villagers, that some people doing well helped others to do so too. He suggests his success is merely an illustration of 'socialism with Chinese characteristics'. However when asked why it was that in Xiao Zhang Zhuang, most households remained tilling the soil, if only on a part-time basis, while in Xie Qiao there were so many successful businessmen, he suggests it was because farmers in Xiao Zhang Zhuang did not have enough money, enough *guanxi* or enough ability. He suggests that things are beginning to change in Xiao Zhang Zhuang, however.

In 1995 both Zhang Jiashun and Su Congfu argued that the future for Xiao Zhang Zhuang lay with the reestablishment of the *jiti* in control of agriculture and that its reestablishment was not merely desirable but *inevitable*. According to Zhang Jiashun, 'socialism can only be achieved through the path of the *jiti*'. He suggested that as farmers progressively gave up the land, the *jiti* would reestablish larger plots, while at the same time allocating people into different economic activities. He said that this was not just *his* opinion but the opinion of all local leaders and leaders at provincial and national level. When asked whether the division of the land in the first place had been a mistake, he suggested that the *jiti* was not strong enough in most places and that land division had aroused the peasants' enthusiasm. However, the reestablishment of the power of the *jiti* would be difficult to achieve where private production was very entrenched. By 1997, Zhang Jiashun was much less bullish about the prospects of *jiti* reestablishing control over the land, indeed, explained that he was now in favour of concentrating land in the hands of capable farmers to create *zhongtian dahu* (land tycoons) as farmers gave up the land to work in enterprises. It is difficult to explain why he had apparently changed his mind without reference to the *relative* lack of momentum of Xiao Zhang Zhuang's industrial enterprises. Though more industrially developed and thus richer than many neighbouring villages in Northern Anhui, Xiao Zhang Zhuang has not reached the same levels of material development that some other villages researched in this work (viz. Liu Min Ying, Doudian, Teng Tou, Qian Wei) and the confidence in the *jiti's* ability to restablish management over the land successfully may well have suffered as a result. Xiao Zhang Zhuang is not yet a *village conglomerate* (see Chapter 9).

Thus, the successful application of CEA in Xiao Zhang Zhuang appears to at the crossroads: the economy continues to expand and the village still applies the 'no-polluting industries rule' of Xie Qiao township, although the thriving brickworks hardly improves the natural environment, the headstocks of the nearby coalmine are clearly visible from the west of the village and the chemical fertiliser factory suggests some weakening of environmental principles. There seems to be a long way to go before the decline of Xiao Zhang Zhuang as *danwei* is witnessed: the apparent level of social solidarity is high. Zhang Jiashun argues that he wants to see Xiao Zhang Zhuang maintain its 'number one' position and spread its influence throughout the Chinese countryside. The village leaders also express their determination to continue development along the same lines, attracting foreign investment if possible, but avoiding polluting industries. Their ambitions are to get still richer, to enjoy perhaps annual per capita incomes by the year 2000 sufficient to allow everyone to have a beautiful home and a car. In 1993 the Fuyang district's 8th five-year plan took Xiao Zhang Zhuang as the model of development. Visitors are encouraged to go and see for themselves and the village has had as many as 700 in one day. With the building of the pleasure park and new zoo, it will continue to attract visitors But for how long it will be a model for CEA, particularly given the *de facto* rejection of biogas, is seriously in doubt.

8(3) He Heng

8(3) 1. Introduction

He Heng is, in 1997, a village of 1969 inhabitants living in 486 households, divided into 7 natural groupings (the old *xiaodui*). Situated in the southern, central region of Jiangsu province, in Tai County, approximately 150 kilometres north-east of Nanjing and 50 kilometres north of the Yangtse River *(Chang Jiang)*, it has 2830 *mu* of arable land and 450 *mu* of water surface. Jiangsu is one of China's richest provinces, its economy fueled by rapid industrial development in the south which has helped to make Nanjing, Jiangsu's 'capital' into one of China's most prosperous and metropolitan cities. The road from Nanjing to Tai County bears witness to rapid industrial development and is an area where joint-venture operations, often with foreign capital are common. He Heng itself is located close to a new canal, the Tong Yang canal, in the Li Xia He district of the Yellow River *(Huai He)* system. The surrounding land is low-lying in relief, being only 1-2 metres above sea level, and is criss-crossed with waterways. It is in an area of China, being in the lower reaches of the Yellow and Yangtse rivers, prone to serious flooding, the last serious flood to affect the area being that of 1991.

He Heng is 10 kilometres from Tai Xian (also known as Jiangyan City), the county town, a busy, industrial town with evident pollution caused by its many chimneys belching smoke and other chemical particulates into the atmosphere. In 1993, He Heng was linked to Tai Xian by only a dirt road which crossed several canals and rivers and which became impassable after heavy summer rains. Its location is pretty isolated, therefore, surrounded, as it is, by fields on the one hand and waterways on the other. In 1995, the dirt road was replaced with a properly constructed modern road, and during its construction there was temporarily no access to He Heng by land at all. My second visit was therefore made by boat from Tai Xian. The rivers are lined with households devoted to fishing living alongside ponds cut out of the river bank and guarded with nets. According to Zhang Tailin of the Tai Xian Environmental Protection Bureau, fishing households are richer than those working the fields, but not as rich as factory leaders, businessmen or even those engaged in transport, plying the rivers and canals with their barges loaded with bricks, coal, and crops. Along the river banks are brick kilns and a building materials factory, all belching out smoke: an all too common sight in present day rural China. By 1996, the new tarmaced road from Tai Xian to He Heng was completed; it is nevertheless still off the beaten track.

8(3) 2. Recent History

Communist forces arrived in central Jiangsu in 1938 and Tia County became a base for the 4th route army in 1945. With the withdrawal of the forces of the *Guomindang* to the south in 1948, He Heng was officially liberated, its three absentee landlords overthrown and its land divided up on a per capita basis. At liberation, He Heng was a very poor village, one of the poorest in the county, its agriculture crippled by constant flooding and waterlogging. By 1964, it had become the richest in the county and has remained so ever since.

According to Zhang Baocun, village leader and Party secretary in 1993, there were essentially three distinct periods involved in the early transformation of the village: the early 1950s through to 1964, 1964-1972 and 1972-80 (its further transformation into an ecological agricultural model village taking place in the mid-1980s). The first period, facilitated, after 1954 by the development of mutual aid teams *(hu zu)* and later by the increasingly cooperative nature of ownership during the 1950s, was characterized by shifts of production away from waterlogged land to dry land. At the beginning of the period,whenever it rained appreciably, and it did so nine years out of every ten, flooding occurred and because of the flat and low-lying nature of the land, it remained waterlogged for long periods of time. Such was the difficulty of tilling the land under these circumstances and the extent of the poverty consequent on that difficulty that some degree of outward migration to other nearby counties to the north of He Heng took place in order to plant dry crops such as wheat. (According to Zhang Baocun, almost all those who migrated at that time have since applied to return and have been refused permission to do so.) In terms of the economic development of He Heng, the critical decision to be made during this period was the decision to plant two crops rather than one crop a year, either two rice or one wheat and one rice, the only village in the county to do so. Political movements of the 1950s had no adverse impact on economic development. He Heng, a brigade within the commune of Shengao by 1958, had a communal dining hall from 1958 to 1962, but the period of the Great Leap Forward, which so adversely affected many other rural areas, appears to have left He Heng relatively unscathed. There was no serious famine in the area.

The period 1964 to 1972 was characterized by an immense programme of works designed to transform the state of the land from waterlogged to dry land. Extensive systems of ditches were dug, dams were constructed, irrigation and drainage systems undertaken and the farmland was built up using the earth dug from the ditches and levelled. In 6 years, 1.8m cubic metres of earthwork had created, from often previously waterlogged land, fields of high and stable yielding farmland. On top of this, decisions were made to farm even more intensively and plant 3 rather than 2 crops a year. Fields were planted with 2 rice and one wheat crop or one cotton and one wheat. Thus, while in the 1950's, yields averaged 100 kilograms per *mu* from one annual harvest, by 1972 yields had expanded to 400 kilograms per *mu* of rice, 400 kilograms per *mu* of wheat and 100 kilograms of cotton per *mu* each harvest. It was this period which coincided with the promulgation of the Cultural Revolution, a movement which led in many areas to the neglect of economic improvements in favour of heightened revolutionary consciousness. According to the current leadership, this did not happen in He Heng because there they followed the teachings of Chairman Mao whose slogan 'catch revolution, promote production' *(zhua gemin ti shengchan),* was interpreted to mean first get something useful done, *then* do revolution.

The period 1972 to 1979 was characterized by further expansion of the system of ditches and dams, afforestation alongside the ditches and a major project to improve the fertility of the soil which was lacking in nutrients, particularly calcium, and of generally poor quality.

Villagers 'worked really hard' to dig sediment out of rivers from an area up to 8 kilometres radius from He Heng, even in the winter months, transferring the mud by boat to the farmland and thus increasing the organic matter in the soil and further raising its relief above sea level. Soils with algae and high levels of calcium were dug in and nitrogen fixing plants were introduced. Thus within three to four years the quality of the soil was transformed, further enhancing the productivity of the land and boosting harvests. Biogas pits were first constructed during this period. At the start of the 1980s, He Heng had consolidated itself as one of the richest and most successful brigades in the county.

8(3) 3. Land Reform

With the onset of the agricultural reforms begun at the end of 1978, the brigade was faced with some important decisions. Despite the high level of trust and respect for the leadership and the extant good relations, there was no opposition in principle to the division of the brigade property amongst, in the first instance in 1982, smaller work teams and, in 1984, households, and as a result, the cooperative *(jiti)* was abandoned, the household responsibility system was introduced and the land divided up very smoothly. The smoothness of the transition, it is claimed today, resulted from the ability and farsightedness of the brigade council who, by the skilful use of propaganda, kept things together. The brigade *(dadui)* became a village *(cun)* in 1983. The standard plot of land was 11 *mu* which was originally divided up amongst two,three or four households depending upon the size of the household. Division amongst individual households took place in 1984. Land was divided amongst households in two forms, contract land *(kouliangtian)* on which farmers had to grow designated crops to fulfil the state contract and private plots on *(zirentian)* which they could grow anything they liked for their household needs. *Kouliangtian* was distributed on the basis of 1.4 mu per head of household, *zirentian* was on average 5.5 *fen* per head per household. There were worries amongst the leaders and some farmers that problems would arise once each household was responsible for production. As a result the leaders took a series of measures designed to encourage and ensure best practice involving such issues as seeding, irrigation, plant protection and cultivation and the management of machinery, in order to allay such fears. Moreover, while the land was eventually fully distributed to households, the village still kept a number of enterprises and assets in collective ownership and management, including certain heavy farm machinery and the growing number of factories and workshops .

Todays leaders defend the decisions that were made to divide up the land and argue that the land reforms of the late 70s and early 80s had a beneficial impact on incentives and thus on productivity and output, grain harvests rising to 1000 kgs. per *mu* per annum by 1985. More importantly, it is argued, they freed the labour force to move to other areas of work. It was from 1978 onwards that village enterprises began to expand. The first such enterprise in He Heng, a factory producing cotton oil, opened as early as 1978, a valve factory opening up soon afterwards. In the early 1980s, further enterprises opened up including an iron boat manufacturing and repairing workshop, a building materials factory and an edible mushroom factory.

8(3) 4. Ecological Agriculture

What environmental advances took place in He Heng before 1984 were not self-consciously 'environmental', rather they were planned and executed purely on the basis of economic expediency. Improvements in soil quality, the construction of dams and ditches, the building up and levelling of land and the planting of trees all come into this category. From 1974 and in line with nationwide trends, a programme of building household level biogas pits was embarked upon, but because of technical difficulties and because of a lack of inputs, there being little animal husbandry and specifically not enough pigs to produce the necessary night soil, the biogas pits were not particularly successful and fell into disuse. However, in 1984, the village leaders, with the full backing of the villagers, made a deliberate decision to take a more self-consciously environmental road.

In the early 1980s, the National Environmental Protection Agency gave increasing publicity to the concept of ecological agriculture and both Professor Li Zhengfang of the Nanjing Research Institute for Environmental Science and officials of the Tai County Environmental Protection Bureau were casting around for a village with an agricultural base which would volunteer to be an experimental 'eco-village'. Indeed, the Nanjing Research Institute had already begun experimentation in another Jiangsu village but a series of disagreements there led to their officials being 'chased out'. He Heng volunteered to take its place.

While the land reforms were popular at the time and, in general, remain defended by the present leadership, it is clear that their implications were not all positive. It seems that the critical factors encouraging the decision to become an experimental village involved certain negative effects of the reforms on the environment: In He Heng the trees were 'privatised'. As a result, in 1984 all the trees which had been collectively planted alongside the ditches and on the edges of the fields were divided up and distributed to households. It was not long before they were cut down and very soon the village and its lands were without tree cover again. Other potential environmentally damaging impacts occurred: while the productivity of the land per se increased, and indeed of labour, as yields were increased and the input of labour fell (peasants increasingly finding employment in enterprises), the process was accompanied by a significant increase in the use of chemical fertilisers and pesticides. By 1983, the average application of chemical fertiliser, mainly nitrogen, to the fields was 100 kgs. per *mu*, 120 per cent more than had been applied before the reforms. This caused a measurable reduction in the organic content of the soil. And as the use of chemical fertilisers and pesticides increased, so did their cost, thus squeezing farmers' incomes.

The leadership was very concerned to remedy this potentially parlous situation and were thus receptive to any ideas as to how to overcome these problems. The opportunity afforded by the Nanjing Research Institute and the Tai County Environmental Protection Bureau to become an experimental 'eco-village' was too good to miss and in consequence the leadership made clear their interest to the county. In 1984, He Heng was accordingly selected to become a model village. Subsequently, a team of technical staff, including Wang Qiuhua, of the Nanjing Research Institute, arrived in the village to begin ecological agricultural construction, staying in the village off and on from 1985 to 1989.

There was no opposition to this development within the village. The leader of the village during my first visit to He Heng in 1993, Zhang Baocun, was at the time working in the county agricultural bureau and took a particular interest in the question of soil and it was he that helped mount, during 1981-3, a series of investigations into the local soil quality and soil nutrients which found that while nitrogen deficiency was not a serious problem, a lack of potassium and phosphate nutrients was, threatening plant and crop productivity. Before the reforms, peasants had organized labour to travel long distances to obtain rich mud from the river, but with land distribution, this became more difficult to do. As a result it became a priority to improve the potassium and phosphate content of the soil and to obviate the need for increased dependency on chemical fertilisers.

The first decision of the Nanjing Research Institute was to construct an ecological demonstration farm, based around pig raising and new, concrete small-scale biogas digesters to provide power for it. (The latter were never very successful: they were plagued with technical difficulties and never fully came on stream). At the same time, the decision was made to convince peasants of the need to raise pigs and to build new household level biogas pits, the night soil of the pigs to provide the inputs for the biogas pits, the outputs of the biogas pits to provide rich organic fertiliser for the fields. More than 200 household level biogas pits were built to provide gas for 550 households. The leaders issued regulations to ensure the successful implementation of the project: peasants had to guarantee that they would provide 100 kilograms of pig manure per annum for use in the biogas pits. Zhang Baocun suggested that the decision to construct new biogas pits was pivotal in the whole ecological agricultural project, since it obligated peasants to engage in animal husbandry, to raise chickens and pigs and produce straw. At the same time as the development of biogas was encouraged, energy saving stoves were introduced to villagers' kitchens.

Other initiatives embarked upon as a result of advice from the technical officers of the Nanjing Institute included the planting of trees alongside the edges of fields, paths and waterways as well as the building up of forest belts at the extremities of the village land. Various ecological agricultural techniques such as stereo cultivation, multi-layer and intercropping were introduced. Other developments were embarked upon to extend and reinforce a virtuous ecological cycle. Fish ponds were dug and the more liquid element of the biogas waste was used as fish feed. Subsequently, the mud in the fishponds was dredged to provide organic fertiliser in the fields. Other forms of cultivation were begun, including mushroom cultivation which was fertilized by biogas slurry, and the breeding of earthworms which ate green foliage from the fields and were used as chicken feed, their waste being recycled back to the fields.

As in many very poor areas of the Third World, there could be little distinction between environmental and economic benefit in the early stages of He Heng's development. Early economic development was largely *predicated* on the environmental changes that took place in the 1950s and 60s, including the digging of ditches, the raising of the fields above sea level, the

142

building of efficient irrigation systems, the growing of trees and the improvements in soil fertility, the classic win-win scenario (World Bank,1992, p.1). However, the effective privatization of agriculture implied by the reforms of the early 80's threatened continued environmental quality in the long run, given the destruction of the trees and the massive expansion in the use of chemical fertilisers and pesticides.

By the end of the 1980s, as a result of ecological agricultural construction, considerable environmental achievements could be claimed. Whereas the proportion of the land covered in forest was, in 1985, only 3.4 per cent this figure had, by 1989 grown to 9.9 per cent and by 1993 to 12.9 per cent, with all the attendant benefits to the quality of the soil and the aesthetic character of the environment. The use of chemical fertilisers was reduced between 1984 and 1989 by 30 per cent as the input of organic matter by way of biogas slurry, mud from fishponds and plant straw was increased. What chemical fertilizer was still used was more carefully monitored and tied closely to the type and quality of soil, reducing the likelihood of future environmental degradation and the cost of fertilizers. Pesticide use between 1984 and 1989 fell by 11 per cent and there were some attempts to develop organic insect control, raising some insects to kill other insects and breeding frogs and toads in ditches to kill other pests. Meanwhile the digging of biogas pits in the 1980s meant that in 1989 biogas provided power for 67 per cent of the cooking and lighting needs of households in He Heng.

8(4) 5. Political Economy in the 1980s and 1990s

8(3) 5.1 Political structures

In 1987, Zhang Baocun, previously head of one of the seven *xiaodui* , was elected *shuji* and leader *(cunzhang)* of the village, a position he held until 1994. However, his official title was changed in 1992 when He Heng village converted itself into a company *(gongsi)* sporting the name 'He Heng Industrial General Corporation' with himself as General Manager and Chairman of the Board *(zhong)*. The company was divided into separate divisions responsible for the various industrial and commercial enterprises, with one responsible for ecological agricultural development. The change in his title, Zhang Baocun claimed in 1993, made little difference to the nature of his responsibilities, suggesting that throughout the time he was leader, the Communist Party has been held responsible not only for political development but economic development as well. In early 1994, Zhang Baocun was promoted to (I was told) other more important work as a manager of a Sheng Gao township and in 1997 he was running an enterprise there producing facial cream.

In 1994, the He Heng General Industrial Corporation appears to have been disbanded as such and Chen Lihua, a 51 year old doctor was chosen to replace Zhang Baocun as *shuji*. He was originally reluctant to assume leadership because he felt he was too old but he was persuaded of a shortage of potential younger leaders and took it on because the Party 'believed in him and asked him to'. At that time the leading group, elected every three years by representatives of villagers, was composed of nine people including himself, the *cunzhang* and accountant, Zhang Zitao, and leaders responsible for agriculture, animal husbandry, social security, women, public security and the Party (youth) League. Formal meetings did not take place regularly, only when important decisions like road-building or the raising of money for a new project needed to be taken.

In 1996 at the age of 28, Zhang Zitao, was elected *shuji*. When asked in 1997 whether he enjoyed being party secretary at such a young age, he replied with a wry smile that he was happy 'to serve the people heart and soul'. Chen Zhangbin, twenty years his senior, became *cunzhang*. In 1997, the number of people on the village committee had been reduced to six, with individuals having more than one portfolio.

It is clear from Zhang Baocun's, Chen Lihua's and Zhang Zitao's testimonies that the Party branch remain influential over decision making in He Heng, although, they claim, not any more so than most other villages. However it is significant that although the land was contracted out at reform the leading group makes the key decisions with regard to land use, specifically what crops are grown, as well as providing advice and technical assistance. And in 1994, the leading group changed the agricultural structure of the village.

8(3) 5.2 Economic developments

The adoption of CEA and attendant environmental benefits would have been given little credence were it not for the fact that He Heng managed to maintain a preeminent position as one of the richer villages in central Jiangsu province, itself a relatively affluent region of rural China. As with many other parts of the Chinese countryside, the expansion of the economy owes much to industrial development provided by sideline enterprises. In the case of He Heng this development, which had begun in the early 1980s accelerated later and was sealed in 1992 with the establishment, between the He Heng Industrial General Corporation and a Hong Kong based company of a joint venture enterprise producing cotton oil for cooking. The advantage for He Heng was that the foreign company put in $1.3m (roughly 13m. yuan) and promised guaranteed sales outside China. The development of industrial employment in this and other enterprises of the Corporation soaked up labour displaced from the fields.

By 1994 although every family still had a small amount of farmland *(sirentian)* only 24 of the 550 households still relied solely on farming. Most villagers worked in the factories at some stage during the year, when there was a lull in the fields, 450 worked in the various factories in He Heng and Sheng Gao, another 300 in transport and other businesses related to the river and 40 were full time fisher-folk. Additionally, independent specialized households *(geti hu)* were responsible for a variety of different forms of raising, including crabs, ducks, turtles and oysters (for pearls).

In 1994, the structure of agriculture was fundamentally changed. After a series of village meetings, at the suggestion of and with the active encouragement of the village leading group, 200 households *voluntarily* cooperated in the construction of two new *jiti* farms, one responsible for animal husbandry, with eight separate units, the other for the production of 'green' rice and vegetables, with twelve units. Most of these households simply gave up their

144

contract land *(kouliangtian)* to work full-time in other enterprises, fishing or transport (keeping only their private land - *zirentian)* t hey were relieved of their obligation to produce grain quota, and were asked only to pay a token tax (of 42.6 *mao* per *mu)* on their private plots. Those keeping their private land were obligated to pay a small annual tax (in 1995, of 25 yuan per *mu)* to the county and fulfil the state contract of 100 jin of rice per *mu.*. The land was consolidated and reallocated, along with the responsibility both for planting and animal husbandry, amongst 20 specialised households. Altogether, 380 *mu* of land was involved. The logic of the exercise was to concentrate what had become increasingly neglected land into new, larger units at the same time as relieving villagers of land they were underusing, thus boosting output and income. In 1997 not only were plots consolidated into larger farms but the agricultural processes had been mechanized: there were two harvesters, one large and three small tractors, a machine for planting, another for drainage and four for spraying ecological pesticide. The green rice produced made a profit of 150,000 yuan. Meanwhile animal husbandry was extensively developed, with 60 breeding pigs, 3,000 pigs for pork, 30,000 chickens and a duck farm, making 70,000 yuan profit.

Between 1994 and 1997 the village opened a small machine tools factory and expanded the boat factory. However, a boost to the economy of He Heng came in 1996 with the establishment of the Green Food Co. Ltd. by Sheng Gao township in cooperation with a number of villages on He Heng's outskirts, processing 'green' rice, making preserved eggs, 'green' pickle, and other delicacies. Zhang Jitao, *shu ji* in He Heng, is a member of the Board and the factory employs twenty workers from He Heng. In 1983, when the greater part of village income was earned from agriculture, the net income per capita was 280 *yuan.* In 1993, with approximately 50 per cent of income attributable to sideline enterprises, net income per head was 1450 *yuan,* the average per capita rural income in Tai County being 900 *yuan.* By 1994, average income per head was 'over 2000 yuan', in 1995 2,580 yuan and 1996 2,800 yuan.

8(3) 5.3 'Social' developments

The expansion of industrial and agricultural output and associated income has allowed considerable improvements in the social infrastructure to take place. As in many other parts of rural China, this is most clearly illustrated in respect of housing. In the 1960s houses were built of mud and straw, in the 70s, clay bricks were used on the outside with large clay tiles on the roofs and mud on the inside, while by the 80s, houses were built with brick on the inside and out, with small, expensively tiled roofs. By the early 1990s there are already enough decent houses for every household in the village. There are relatively few 2-storey houses- only 10 per cent of the total stock- and the traditional courtyard style house remains popular. In 1996 a new housing drive was underway however: 58 two-storey houses were built in the middle of the village close to factory sites, occupied by villagers working in enterprises who had returned their land to the collective. The houses are built with solar panels in the roof, are amenable to biogas, and all have gardens planted with flowers.

Educational opportunities have expanded considerably in recent years: there is now a kindergarten and a primary school with capacity for up to 300 children in the village paid for from village funds and the proportion of young people attending middle school is almost 100 per cent. Few did so even as late as the 1970s. Many students go on to attend professional high school and, since 1978, 70 have attended university (none of whom have returned to live in the village). There is a small hospital and a recreational facility for villagers, with table tennis and karaoke, also paid for out of village funds, but there is no old peoples home: those old people without relatives to care for them are sent to Sheng Gao, the nearby township.

Social development is clearly charted by the leap that has taken place in recent years in material expectations. In the early 1980s when young girls got married, while they would be adequately fed, they would perceive a black-and-white television to be a luxury, their clothing would be made of cheap artificial fibres and they would be lucky to get a brick and tile house. By the late 1980s, according to Zhang Baocun, this had all changed. Now when they got married, they could afford all sorts of electrical appliances, including a colour television, refrigerator and video recorder and they would live in brick-built 3 room house costing 30-40,000 yuan.

According to the leadership, He Heng has maintained a very strict birth control policy, based on the nationally promulgated one-child policy. Only those families allowed under the provisions of the policy to have 2 children (e.g. those whose first child is disabled) have done so. It is argued that the policy needs little heavy enforcing: the second child which up until recently was perceived to be an economic boon is in the 1990s more likely to be seen as an economic burden. Zhang Tailin did say, however, that should a second conception occur, the mother would be expected to have an abortion and, if she did not, she would be fined. Life expectancy is high and continues to rise. Zhang Tailin suggested that not so long ago, there was frequent disease as people drank dirty water and suffered from the effects of applying chemical fertiliser and pesticides. Nowadays, he suggested 'the environment is getting better and people are living longer.'

8(3) 5.3.1 Personal experiences

(i) In 1993, Zhang Baocun was 41 years old and *shuji* of He Heng village, having been elected to that position in 1987. He married Chen Kezhen, two years his junior, in 1975. Zhang Hongmei was born in 1978. They lived in a single-storey, 5 roomed traditional walled courtyard home, built in 1980. Their home had tap water, a range of consumer durables and a huge poster of Mao Zedong on the living room wall. The courtyard was bedecked with flowers and the walls with grapevines. They were keenly aware of the economic changes that have taken place in their lifetimes and enjoyed their new-found affluence, arguing that they had become 'careless' about money concerning small things and only cared about it when big things needed to be bought. Chen Kezhen was the eldest of 7 children and had had no schooling, having to help her parents with housework as well as work in the fields. She was born during the period in which the flooded land was converted to flat land and her lack of schooling was not a unique

146

experience. She and her husband were therefore very much aware of the progress that had been made in recent times and were pleasantly envious of their daughter's educational opportunities. In 1993, Zhang Hongmei was in the third grade of middle school, progressing very well with parental ambitions for her future career involving either something in a medical capacity such as a doctor or nurse, or accountancy, both of which would involve her attending a professional high school in the years ahead. Since they were concerned that political change may take place at some stage in the near future (meaning that the reforms might be reversed), Zhang Baocun felt the medical profession would be safer, particularly for a girl. Zhang Hongmei's school paid considerable attention to 'environmental propaganda', teaching environmental protection in geography classes. He Heng was seen to be a model for environmental advancement and the school arranged visits of students and teachers there.

(ii) In 1993 Chen Lichuan was 60 years old and lived with his 58 year-old wife in a five roomed courtyard house. Their family was originally five strong but their two daughters were married and lived with their husbands and their son had left for university and had not returned to the village. Chen Lichuan lived in He Heng at liberation and remembered living conditions as being very poor at that time. He was the elder brother of two, neither of whom went to school (nor did his present wife). They had lived in a one-roomed house of mud and straw and they never ate meat or vegetables, only a porridge made from rice. His parents worked very hard in the fields and he did not see much of them when he was young since they left the house at 7am and rarely came home before 7pm. He married his wife in 1958 and ate whatever foods were available in the communal mess hall in the first years of their married life. He suggested the two critical years for He Heng's recent development were 1973, when the decision to grow three crops rather than two was made and 1984, when land distribution took place. His present home was one of the newest in the village, built in 1985. It was as late as 1962 that he and wife first lived in a two-roomed house and then the mid-1970s before his mud house was reconstructed in brick and tile. Chen Lichuan fully supported land distribution (and remained supportive) and was equally supportive of the ecological agricultural developments. He could sell his vegetables in the market and used biogas slurry on his fields. Owing to his age, he worked only in the fields.

(iii) In 1993 Chen Lijiang was 50 years old and lived with his wife also in a 1980s built single story five-roomed courtyard house. They had three children, two sons, one of whom graduated from university in 1993 and one married daughter. Chen Lijiang and his non-graduate son and daughter all worked in the same factory in He Heng. He and his wife have only four *mu* of land on which he planted rice and cotton, using organic fertiliser, spending half his time in the fields and the other half in the factory. He put the economic success of the village down to three things, government policy, -specifically land distribution -strong village leadership and ecological agriculture. Land distribution he saw to have been of importance because it allowed more peasants to work in the village enterprises. Chen Lijiang had biogas for cooking and lighting and was very happy with and proud of it. (I was given a demonstration at his home). He argued biogas would remain very important for many years to come: it was hygienic and non-odorous in the home and provided nutritious slurry for the fields. He suggested that compared with other villages, the leadership had been strong and that relationships *(guanxi)* between people in the village had been and currently were, for the most part, very good. He argued that the present and future success of biogas would depend very much on government attitudes at the regional, provincial and national levels, suggesting that the government had a duty to convince people of its value. He agreed with Zhang Baocun that the attitude of the government would be more important than the price of electricity. He saw the need for an extension of biogas into the cities as well as the countryside and perceived the issue as a world problem.

8(3) 5.4 Environmental developments

He Heng was nominated by NEPA for the UNEP Global 500 'role of honour for environmental achievement' award and became China's 11th recipient of the honour in 1990. The citation read:

> Traditionally, this village was poor because it was low-lying and flooded annually. Following a programme which includes afforestation, constructing marsh biogas ponds for fuel, establishment of an eco-farm, it has been transformed into a flourishing and prosperous area. (UNEP, 1995)

He Heng was the third Chinese village to win the award for developments in ecological agriculture. The receipt of the award was also a relatively low key affair - no foreign trips for Zhang Baocun nor official celebrations in the village.

While He Heng successfully adopted CEA in the 1980s, its position as a model of ecological agricultural construction was being questioned by the mid 1990s as a result of several factors: the encroachment of polluting industry into the economy of the village, the fall in biogas production and thus of its use in households for cooking and lighting, and, partly because of the latter, a heavier application of chemical fertilisers than was desirable. All of the above were observable. In 1993, the cotton oil factory was causing atmospheric pollution, by 1995, it was admitted by the leaders that biogas was enjoyed by only 20 per cent of households, a long way from the 70 per cent claimed at the time of UNEP award, with 80 per cent of households still using stalks as their primary fuel and with the consequent loss of organic fertiliser. By 1996, biogas generation was negligible: 'only five or six houses still use it', I was told.

Possible explanations for this regression include the fact the He Heng's leadership has not been as stable as in many other eco-villages: Zhang Baocun became leader only in 1987 and left the job in early 1994, since when there have been two further *shuji*. As a result, it is likely that the continuity - and obstinacy - necessary to pursue a particular path, particularly one where short-term payoffs may not be immediately obvious, was to some extent damaged. The continued progress in ecological agricultural construction would not have been helped by the withdrawal of the team from the Nanjing Research Institute and a substantial reduction in the ongoing technical support provided by them after 1989. The concrete biogas pits built on the demonstration farm built in 1986 never overcame technical difficulties (they were used for fish in 1995). And the isolated location of He Heng would have put it at a disadvantage in the general scramble to become rich in the 1990s, making it more difficult to say no to industrial development, even when, as in the case of the cotton plant, there was the threat of pollution. The collective is weaker in He Heng than in some other eco-villages, where the land is still communally farmed.

Moreover, in 1996, the optimism expressed about biogas in He Heng only two or three years earlier had been dissipated. Bian Dongan, deputy chief of the Environmental Protection Bureau in Tai Xian, explained the decline of biogas as an inevitable process, associated with the farmers getting richer. He suggested that biogas had been originally encouraged in China in the 1970s primarily as a result of energy shortages and had had nothing *per se* to do with environmental protection. Farmers had constructed individual household biogas pits to provide power for cooking and lighting, but it had not been very successful and involved very tiring, time-consuming work. By 1996, the farmers were a lot richer, were supplied with electricity and gas and did not want to bother with biogas even if it was now associated with environmental protection. For Bian Dongan, biogas at the household level was a thing of the past, associated with poverty and a lower level of rural development: the farmers were simply taking a very pragmatic view. This accounted for the continued application of chemical fertilisers as well, he explained. When asked about the possibility of building a communal level biogas digester, he was very pessimistic, arguing that He Heng (and other similar villages) could not afford the 800,000 yuan initial capital investment necessary and would not be able to overcome the problems of input collection, maintenance and distribution to all households. In 1997, the leaders were even more entrenched in their views, arguing that while originally biogas had provided organic fertiliser in short supply and had 'liberated' women in the kitchen, now it was not 'user-friendly'. Villages could burn cylinder gas for cooking and had electricity for lighting, animal husbandry meant excrement could be used directly on the fields while the household-level pits were too troublesome (*tai duo mafan*) to build, seal, fill with excrement and other materials as was the application of the slurry on the fields. Zhang Tailin suggested that a community-level biogas digester might be built if He Heng had the money but it doesn't.

However, despite an obvious change in its nature, by 1996 and 1997 there was evidence to suggest that ecological agriculture in He Heng was showing some resilience, particularly with the growing of 'green' food. The decision to do so was, according to Zhang Tailin, made for three reasons: (1) it was healthy (people wanted a better life), (2) the conditions were suitable in that there was sufficient organic fertiliser produced in the various animal husbandry units and (3) it was profitable - green rice could net farmers 2.6 yuan per *jin,* ordinary rice only half that much. The decision to grow it in He Heng was because of its previous reputation and tradition as an eco-village.

In 1995, Zhang Tailin explained that He Heng's plan was to become a model village once more on the basis of the application of ecologically scientific principles to production, including planting, animal husbandry and industry. With regard to the former, He Heng was chosen in 1994 to be an experimental site for the growing of 'organic' and 'green' foods by Prof. Li Zhengfang's team at the Nanjing Research Institute hence the consolidation fields of the fields in 1994, growing green rice and experimenting in other organic products including organic cotton.

149

The green food farm, run by the *jiti* is worked by 22 villagers, categorized as *nongye gongren* (agricultural workers) each earning approximately 4,000 yuan in 1996. By 1996, the lush 'green' rice fields stretched as far as the eye could see, behind a huge billboard, advertizing 'GREEN FOOD' in Chinese and English. As with all planting, irrigation is the responsibility of the village as a collective, the money coming from enterprize surpluses. Xiao Xingji, now director of NEPA Organic Food Development Centre at the Nanjing Research Institute, expressed cautious optimism over the likely success of these experiments in He Heng. The 'green' rice can be sold for up to 3 yuan per *jin* (more than double the 1.3 yuan per *jin* for ordinary rice), ending up in hotels and big cities in China. It is presently sold through an agency (originally arranged by the Organic Food Development Centre) although there it is hoped that, in near the future, markets abroad, perhaps in Japan and Europe, will eventually be found. In the near future, it is planned to grow organic grapes and introduce organic fruit trees.

With regard to industry, Zhang Tailin reaffirmed He Heng's commitment to non-polluting industry, citing the recent rejections of a zinc plating factory and a potentially very profitable facial cream factory on grounds of pollution.

8(3) 6. Conclusions

In the late summer of 1995, China Daily commented,

> He Heng, a village of Jiangyan City in East China's Jiangsu Province, is expanding. A report by Jiangsu Provincial Environment Monitoring Station showed the water, soil and air in He Heng village are above State standards.
> The village has demonstrated an example of modern agriculture, and is a window showing how China fulfils her obligations to the human beings in the world', said Jiang Chunyun, vice-premier of the State Council, during a visit....
> Last year the village's grain output was 627,000 kilograms and per capita'annual income was 1,789 yuan. The food was labelled 'green food' or non-polluted food by the state and sold for double that of common kinds of grain. The village spent 5 years cooperating with an environmental protection research institute and turned 147 hectares of land into a protected area divided by green tree belts into high yield blocks, with an underground irrigation system.
> The use of fertilisers and pesticides is strictly controlled. Every family raises livestock and uses methane as fuel. To curb the air and water pollution, the village shut down a cotton plant and several dozen brick kilns and set up pollution free enterprises (25/8/95, p.5).

While the above paragraph contains a good deal of artistic licence (e.g. in the summer of 1995 only a very small proportion of grain output was 'green', every family did not own livestock and only a handful of households used methane [biogas]), its inclusion in China Daily in late 1995 reflected, if only as a result of high level *guanxi*, He Heng's revived interest in environmental advance and its rehabilitation into the family of model villages.

There can be little doubt that the development of He Heng as an ecological agricultural village owes more to patronage of the Nanjing Research Institute of Environmental Science, and specifically to Professor Li Zhengfang's team of ecologists, including Xiao Xingji and Wang Qiuhua, first in their guise as nature conservation experts during the late 1980s and subsequently in their reincarnation as leaders of the NEPA Organic Food Development Centre in the 1990s, than to anything else. This patronage has gone hand in hand with patronage from the Tai County (Jiangyan City) Environmental Protection Bureau.

The leaders of the village were imaginative and flexible enough to have been open to their advice and technical assistance in the 1980s, the village was relatively rich and had a large accumulation fund and there were high levels of cooperation in the village leading to successful ecological agricultural construction and the winning of the UNEP 'Global 500' award in 1990. But when that assistance was reduced to a minimum shortly afterwards, the village wavered in its commitment to ecological advance and only the new initiatives of the Organic Food Development Centre in Nanjing and the County Environmental Protection Bureau, itself encouraged by its adoption as an 'ecological county' in 1994, renewed that commitment. My strong impression of affairs by the time of my second,third and fourth visits in 1995, 1996 and 1997 was that the Tai County officials were effectively taking the initiative in determining He Heng's destiny; indeed it was made clear that He Heng had been specifically *chosen* as one of 2 villages in the county to be exemplars of ecological agricultural development in 1994, Zhang Tailin suggesting that the reasons for the choice of He Heng being that 'the basics were already better there and the cultural quality was higher there than in other villages.'

Perhaps one other factor explaining He Heng's renewal has been the increasing power of the *jiti* over non-industrial production as a result of the consolidation of the fields and the increased development of animal husbandry in 1994. When asked the relative merits of household or *jiti* farming in 1997, Zhang Tailin said that while individual farms may be better in the short run, collective farms may 'take better care of the land' and have 'a wider view.'

In 1995 Zhang Tailin considered that He Heng had merely been an *experimental* model village in the 1980s and that in the late 1990s it would become a *real* model village, depending on its own resources, engaged in 'low input, high output' production around the planting of 'green' foods, and applying scientific ecological agricultural principles. This development has given the village, as a collective unit a new lease of life: as Zhang Tailin explained, there isn't any point individual households growing organic crops if neighbours on adjoining plots still use chemicals. Now no chemicals at all are used on the collective fields, all the fertiliser comes from the pigs.

Given the continued commitment of his Bureau and the interest and patronage of the Nanjing team, He Heng has been successful in boosting incomes along an essentially 'green' route. Indeed, it is hoped that in a couple of years some of the crops will achieve 'organic' status, In the mid/late 1990s, however, He Heng remains in a state of flux and the outcome is unclear: Zhang Tailin knows all too well that farmers are essentially materialistic and the capacity of He Heng to be an effective exemplar and encourage the spread of ecological agricultural principles will depend on the ability of its villagers to make large amounts of money for themselves. The growing of 'green' rice and its processing in a new 'green' rice factory employing local villagers seems to be commercially viable. On the question of whether further extension of ecologically sound practices will be so in the medium term, without the sponsorship, technical assistance, and the general 'Hawthorne' effect which previous model status has provided, the jury is still out. Certainly, biogas is a thing of the past and the archetypal ecologically virtuous cycle with biogas at its core is not on the agenda.

8(4) Dou Dian

8(4) 1. Introduction

Dou Dian, is a largish village of 4100 people in 1230 households, about 45 kilometres south
west of Beijing and 10 kilometres north of Fangshan, the county town. It has 5,200 *mu* of
fields and with its residential and industrial belts, its total area is approx. 10,000 *mu*. Standing
38m. above sea level on the North China Plain, its surrounding area is flat, it benefits from
underground water and has a sandy soil suitable for crop-growing. With the expansion of the
urban area of Beijing and improvements in road and rail communications, Dou Dian, although
surrounded by fields and still in a rural location, is increasingly being drawn into the suburbs of
the capital: it has for many years had a railway station (some distance from the village centre) on
the Beijing-Guangzhou and Beijing-Shi Jia Zhuang lines, but with the building of the Beijing-
Shi Jia Zhuang motorway running adjacent to the village in the last couple of years (a motorway
planned to stretch all the way to Shenzhen in Guangdong province), it is within easy road
access of Beijing and plenty of motor traffic travels between Dou Dian and Beijing daily. A trip
to Beijing and back for a cadre or, indeed worker in Dou Dian, is increasingly a routine state of
affairs.

Dou Dian is a well-known village (many people in Beijing to whom I spoke casually
about my visits had heard of it), the primary reason, it seems, being that throughout the 1980s
and early 1990s it consistently won awards as a model socialist village while its leader and *shuji*
Zhang Zhenliang won awards as a model worker and party member. In the late 1980s it adopted
CEA and while, in some respects, it has become a 'model' ecological agricultural village,
attracting visitors interested in that process, it attracts many more as one of the richest and
industrially most dynamic villages in the Beijing municipality. Its primary claim to fame, in the
1980s, as now, is that its achievements have been made with the village as a wholly cooperative
organisation: it remains *jiti*, having never divided its fields or property. It is not for nothing that
one of its propaganda sheets suggests that 'Today's Dou Dian is like a bright socialist pearl
which is dazzling and bright.' (Dou Dian, 1994, p.2)

Dominating the skyline of Dou Dian is a huge dome, shaped as though it belonged to a
Muslim temple . (On closer inspection the dome incongruously sits on top of a recently opened
and already scruffy public canteen). The Islamic influence is explained by the large number of
residents (20 per cent) belonging to the Hui minority, *shuji* Zhang Zhenliang himself being a
Hui, all of whom regularly worship in the newly built Muslim temple in a small backstreet.
Other minorities amongst Dou Dian residents include Man and Zhuang.

There is plenty of evidence of recent construction, notably the new Science and Technology
Building, which functions as the main public building of Dou Dian, housing the cadres' offices,
reception rooms, guest rooms, karaoke room and other business and leisure facilities. Opposite,
new two-storey houses have been built. However, despite the existence of a number of new
buildings, Dou Dian does not give the impression of being a neat, model village in most other

respects, indeed rather the opposite. Its wide main street is a major thoroughfare with plenty of traffic and its immediate environment is overtly disorderly and scruffy. Indeed, it is difficult to reconcile the evidence to ones eyes with its claim to be a 'bright socialist pearl.'

8(4) 2. Recent History

Dou Dian has a long history, reputedly having been founded in the Sui dynasty (589-619 AD) by Dou Jiande, the leader of a peasant insurrectionary army, who built a fortress and stationed troops there. But just outside Dou Dian is Dong Jialin village, the oldest archaeological site in the Beijing area and the capital of the Yan State of the estern Zhou dynasty. The area thus has a history stretching back over 4,000 years, and there is a large new museum attracting frequent visitors just a couple of kilometres from Dou Dian.

At Liberation in 1949, Dou Dian had the status of a town *(zhen)* with a total land area of 30,000 *mu* encompassing 12 villages with a total population of about 30,000, and while most villagers were engaged on the land, some of the more powerful people of Dou Dian were businessmen with dealings in Beijing. According to Xu Jianyu, current head of agriculture in Dou Dian, there were four or five landlords who, after Liberation, were subject to reeducation, forced to work alongside others and whose land was taken away and divided up amongst the peasantry, each peasant having one *mu* only. The landlords were not otherwise persecuted at that time although they later suffered badly during the Cultural Revolution. While in 1949, the fields were given to individual peasants, by 1952 small teams of workers farmed together and by 1956, a cooperative *-jiti -* was established in what is now Dou Dian village.

In the 1950s, in common with most parts of the Chinese countryside, life was hard and peasants were poor. Productivity in the fields was low, sometimes as low as 100 jin per *mu,* there was no rice to eat, only maize, often in very short supply and almost all the houses were built of mud, inside and out. The very best houses had bricks outside. The traditional *kang* provided the only opportunity for heating and cooking and was used by the whole family to sleep on. Electricity did not come until 1961/2 and was then only used for lamps. Wood stripped from the branches of trees and stalks from the fields were the primary fuels for cooking and heating although sometimes there was a little coal.

During the Great Leap Forward launched in 1958, Dou Dian *zhen* became a commune *(gongshe),* while one of the twelve villages *(cun)* within it became Dou Dian brigade *(dadui),* itself divided up into eleven production teams *(xiaodui)* the latter becoming the basic accounting units. There was no major steel production in Dou Dian during the Great Leap Forward although a communal dining hall was built. From 1958 to 1961, Dou Dian brigade members all ate in the dining hall, a popular enough institution at first because people could eat as much as they liked. Subsequently, however, owing to shortages of wood for cooking - most trees had already been felled and even some buildings were torn down to get hold of wood - the dining hall couldn't provide enough food which accordingly became rationed, and everyone was issued with food tickets. However, according to Xu Jianyu, despite the privations 'there were strong Communist feelings' buttressed by slogans suggesting that with a Communist society 'you will get a telephone, eat bread and drink milk.' While life was very poor, it was nonetheless 'secure and peaceful'. However true this might have been, Xu Guangkun, the 44 year old leader of the pig factory, remembers 1962 as being a particularly bad time, with food restricted to 4 *liang* of corn per person per day.

In the 1960s, things got little better. All brigade members worked on the fields producing wheat, maize and soya beans. Households dug wells to benefit from underground water, but there was no irrigation on the fields and the success of the harvest depended on rain. No vegetables were grown and while a few households kept a pig, there was often difficulty in finding sufficient fodder to feed it. There were, however, one or two local handicraft industries, notably straw hat making and and the forging of horseshoes.

The Cultural Revolution had a significant impact in Dou Dian: there was very considerable political activity and faction-fighting. Politics was 'in command'. Zhang Zhenliang, who had, at the age of 28 become Party secretary *(shuji)* in 1956, and who had been, at liberation, a middle peasant, suffered alongside ex-landlords and rich peasants as he was struggled against and deposed from office in 1966. Zhang Zhen-liang recalls that 'all leaders had problems during the Cultural Revolution'. According to Xu Jianyu, however, he was brought back as *shuji* two years later because nobody else could be found to do the job well enough. The political movements also had a serious impact on production: in one section of farmlands (no. 13), peasants were so politically active that they did not have time to water the fields and as a result, the wheat crop failed leading to very serious food shortages. At the same time, all non-farm activities, such as pig-rearing, chicken rearing or handicrafts were outlawed and denounced as 'capitalist tails'. According to Xu Guangkun, 'nobody did anything they should have done, they broke everything, there was much political activity'.

The impact of the Cultural Revolution continued to remain strong in Dou Dian only until 1969. Nonetheless, things remained difficult in the 1970s and most household activities other than planting were outlawed. On the other hand, the village, as a cooperative *(jiti)* resumed the production of straw hats which were sold to the county, who in turn sold them in the market. The hats were made with stalks and each worker employed was paid 5 yuan a month. It was one way in which brigade members could earn some extra money. At the same time, agriculture left the villagers poor: the success of the harvest depended totally on the weather, wells were few and irrigation basic. Wheat and maize growing were the dominant activities, the fields were fertilised almost exclusively by human excrement and output remained as low as 700 *jin* per *mu*.. There was very little animal husbandry, the *jiti* owning around 100 head of cattle and horses in total and there was no agricultural machinery other than one tractor. In the 70s there was a market every three to five days where, according the Zhang Zhenliang, 'people sold rubbish.' Vegetable production was negligible. Dou Dian found it difficult to feed itself, let alone provide a surplus for the county.

As in common with other claims made by leaders throughout China, Xu Jianyu argues that things only began to change significantly for the better in Dou Dian in 1978. However, unlike those other claims which predominantly highlight the importance of the decisions made by the 3rd Plenum of the Communist Party Central Committee in December 1978, Xu Jianyu puts as the primary catalyst for change decisions made by the *local* leaders to apply scientific

155

principles to agricultural production. In 1976, *shuji* Zhang Zhen-Liang, realizing the need for education, met with a leader from Beijing Academy of Agricultural and Forestry Sciences [subsequently Beijing Agricultural University] who was then experimenting with scientific agricultural methods in a small brigade in the environs of Beijing, who wanted to extend his project to a bigger one and was interested in Dou Dian brigade because of its larger size. According to Zhang Zhenliang, this was the first time any brigade leader had asked him for help. Accordingly, Cheng Xu, a scientific officer at the Academy was invited to come to Dou Dian with a team of agriculturalists and apply new scientific principles to agricultural activity there. This crucial development was, according to Xu Jianyu, the result of 'the great imagination of *shuji* Zhang.'

Cheng Xu and a team of up to nine workers arrived in Dou Dian in 1979 and stayed off and on in the village for the next ten years. Initially, the innovations were unpopular amongst villagers: they involved changing practices and working very hard. The team's early recommendations seem to have chimed closely with the theory and practice of the 'Green Revolution' popular at that time. According to Cheng and Simpson,

> in effect, the tactic was to expand grain output to (1) meet village human needs, (2) produce a surplus which could be sold to secure cash for inputs and capital investments and (3) expand feed grain output. Livestock became a central focus, both as a means to improve protein availability for villagers and as a contribution to cash earnings. (1989, p.4)

Cheng Xu and his team first attempted to improve the cropping system by introducing double cropping, growing wheat in winter and maize in summer. They gave particular concern to the problem of controlling weeds and killing pests, the former problem dealt with by the double ploughing of the land prior to the planting of winter wheat, the latter by the application of chemicals. Cheng Xu further experimented with 3 new chemical fertilisers, examining soil quality in different fields to determine the optimum blend of fertiliser in each field. New seeds were introduced, including new rice strains from Hainan Island. And machinery was bought with collective funds. Beforehand, the *jiti* had only the one tractor used only by a very few brigade members. Subsequent to Cheng Xu's reforms, several new pieces of machinery were bought, including specialised machines for seeding and harvesting. The reforms had a significant impact on farm productivity and output: in 1977, grain output was 700 jin per *mu*, by 1985, it reached 1390 jin per *mu*, almost double the average yield produced in other parts of China and close to yields experienced in the USA. Maize became a more important crop than wheat. The 1978 harvest alone was so large that Dou Dian provided over 1m. jin of grain to the county. And as agricultural productivity grew and production production expanded, the village which had still been poor in 1978 (a 'slum village', according to Cheng and Simpson, (1989, p.4)) with no communal savings, slowly but surely became richer.

Increases in productivity caused by new techniques and machinery meant not only increased output, but the release of labour from the land: the result was an early decision by the leadership to develop both farm and non-farm enterprises, in particular to develop animal husbandry. The importance of such development was not simply to allow the diversification of

156

the Dou Dian economy per se, but to solve the problems attendant on releasing labour from the fields: According to Cheng and Simpson,

> At first, great resistance was met through belief that heavy human population density and a large labour force precluded mechanisation. In other words, it was felt that employment as a farmer, however unproductive, was better than the only other apparent alternative- being idle. This thinking was overcome by development of village enterprises (1989, p.4).

The first non-industrial enterprise was a chicken farm, set up with the help of the Beijing Academy of Agricultural and Forestry Sciences, which, by 1982, had 30,000 chickens. Next, a pig farm was established and cows bought. As a result, it was possible for Xu and Simpson to report that, by 1987,

> Introduction of livestock into the village as a major production complex is beginning to yield considerable by-products, most notably manure. In 1977, no cattle was being fed and there were only 2,000 layers and 700 hogs. By 1987, the village was fattening about 2,700 head of cattle annually and had 40,000 laying chickens and 2,500 head of hogs (1989, p.5).

While in 1978, there was almost no trees at all in Dou Dian, by 1982, a programme of afforestation began. And by 1983,

> 24 village run enterprises had been developed which produce such varied articles as garments... prefabricated concrete parts, bricks, carpets and handicraft items. (1989, p.4)

Thus by 1983, the economy of Dou Dian brigade was experiencing the process of diversification and enjoying its initial fruits.

8(4) 3. Land Reform

Dou Dian brigade became Dou Dian village *(cun)* in 1982 and in common with all villages in the municipality of Beijing was ordered to abandon its *jiti* organisation, divide its fields and property and introduce the HRS. However, *shuji* Zhang and the rest of the leadership were opposed to doing so. They argued that the economic advances recently made were predicated on communally owned machinery operating on communal fields and that division of the land and property would effectively reverse those advances as private plots became too small to use the machines effectively with the risk that, in the worst case, the machines themselves would be dismantled and split up. As in all recalcitrant units, the Beijing municipal authorities came down to Dou Dian to force the leadership to change its mind and they made life difficult for *shuji* Zhang Zhen-Liang. But unlike in those other units, they were not successful in getting their way. The leadership was supported in a series of meetings by the villagers who wanted to continue the mechanisation and diversification processes, happy with the economic benefits which were accruing. According to Xu Jianyu 'noone wanted to divide the fields, villagers were very satisfied with the high level of mechanisation, they didn't have to work so hard.' And in the end, with the support of the Beijing Academy of Agricultural and Forestry Sciences, the leadership managed to hold out. Dou Dian never divided its fields nor its property and has remained fully operative as a *jiti* to this day. The Beijing authorities do not appear to have held a grudge for long: *shuji* Zhang Zhenliang began winning awards as early as 1982, Dou Dian was praised as a 'model civilised village of Beijing' in 1985 and in 1986, a delegation from the National Government headed by Vice-Premier Chen Yun came to the village, primarily interested in its *jiti* form. It was then that the village first got written about in the national newspapers.

8(4) 4. Ecological Agriculture

Dou Dian is today a renowned village in the Beijing area not primarily because it has adopted CEA. In fact the introduction of ecological agricultural *followed* its initial 'model village' status. Once again, Cheng Xu's interventions seem decisive. While Xu and Simpson are clear that the 'Green Revolution' type developments that Cheng Xu and his team had ushered in in the late 70s and early 80s had been very beneficial in terms of productivity and output, there were nonetheless potentially serious implications which had to be taken on board:

> The question that arose a few years ago was what to concentrate on next - especially given the greatly expanded input use..... It was determined that, while yields had been increased dramatically through the use of chemical fertilisers and a variety of energy-using inputs, China's ecological system would suffer greatly if this micro-level system, apparently a very successful experiment, became practised throughout the arable lands of China.... Dependence on heavy chemical fertiliser application would mean that agriculture would be contributing to an eventual energy crisis.
> Perhaps an even more serious problem is concern about pollution. Heavy input of inorganic energy, especially in chemical fertiliser, combined with a highly intensified animal husbandry, could cause environmental problems such as leaching and enrichment of nitrates in underground and surface water, as well as odour. These problems have become very important recently (1989, p.5).

Xu Jianyu suggests that Zhang Zhenliang and he first heard of the concept of ecological agriculture in 1985 from Cheng Xu. The catalyst for concern was the pollution associated with excess manure from the chicken factory: the chickens' excrement was going straight back to the fields, risking not only surface and underground water pollution, the latter particularly serious because it is not immediately visible, but a form of pollution rather closer to home. Xu Jianyu explained with a huge grin, 'the shit got on the feet of peasants who walked back into the village. It led to a terrible smell.'

Some households had built individual biogas pits for cooking and lighting, many fed directly from household toilets: even Zhang Zhenliang had his own biogas pit in the 1970s. In

158

the mid-1980s, at the suggestion of Cheng Xu, the village introduced a 400 cubic metre biogas digester near the chicken house and cow sheds, big enough to treat manure from 600 head of cattle daily. An ecological agricultural *system* gradually developed from then on. Chicken and cattle manure fed the biogas plant capable of producing 250 cubic metres of biogas a day, initially providing households with clean fuel for cooking, its solid waste being returned as enriched manure to the fields while its liquid slurry, along with poultry manure and waste blood from animal and poultry slaughter operations, was used to feed fish and snails, introduced to the village at the same time to extend the ecological cycle. Snail breeding was introduced on the grounds that snail meat was rich in protein and their shells rich in calcium and magnesium, allowing finely ground snail shells to be used as a partial substitute for imported fish meal. Slowly but surely,

> The lessons of thousands of production cycles in which nature has harmoniously co-existed with humans was drawn upon, albeit in a sub-conscious manner, and a biological cycle farming system devised (Cheng Xu and Simpson, 1989, p.5).

Thus, what today in China is known as CEA became grounded in Dou Dian by the late 1980s. By mid-1988, the biogas plant was operating at 100% capacity, not only providing households with fuel for boiling water and cooking but at the same time powering a 300 KWH generator providing electricity for the entire egg-laying operation. Additionally, it was providing 45 tons of liquid sludge to feed the fish and snails; by mid-1988, 5 tons of fish and 0.1 tons of snails were produced from the 3 hectares of ponds. Meanwhile, the solid sludge was replaced 12-15 per cent of the concentrate used in pig fodder, important given the importance of pork in the Chinese meat diet, and the rest was used as organic fertiliser on the fields.

Other elements of an ecologically beneficial cycle involved the use of poultry manure mixed with crop residues to provide animal fodder and the production of 'so-called fungi-bran feed using corncobs, which previously had little value, as the cultivation substrate of mushrooms. The substrate is fed to hogs after mushrooms have been harvested several times.' Also introduced at the time by Cheng Xu and his team was the growing of:

> azolla, an aquatic plant community consisting of a fern and symbiotic blue-green algae...which fixes its own nitrogen and requires only additional phosphate. In the Dou Dian experiment, it covers the ponds where the snails and fish are grown and is provided with additional phosphate from the liquid sludge. It reproduces quickly by means of highly effective photosynthesis and nitrogen fixation (Cheng Xu and Simpson, 1989, p.6).

It is clear that Dou Dian, has taken a pragmatic approach to organic agriculture and that its current operation is by no means pure. As Cheng Xu, Han and Taylor suggest, the absolute industrialized energy intensity of agricultural production had *increased* up to 1988 and at 23.4 m. kilocalories per hectare remained high, but that:

> overall biomass output has increased considerably and efficiencies of energy and other resource utilization in the village have improved

modestly (1992, pp.1137-8).

Since 1989, further development of ecological agriculture has taken place, albeit rather gradually. In 1991, the *jiti* built a second, smaller biogas digester with a capacity of 100 cubic metres close to the pig factory and fed exclusively from it. This allowed a further 250 households to be provided with gas for cooking free of charge. In the winter of 1994/5, the larger biogas digester had technical problems and was, during my visit in the summer of 1995, out of operation. By 1996, however, both digesters were back in operation and in 1997 a third smaller 30 m^3 biogas digester had been constructed to warm new greenhouses built to grow 'green' vegetables and raise pigs in winter. Dou Dian was embracing the Ministry of Agriculture's latest biogas slogan 'four-in-one': gas, vegetables, animal husbandry and fertiliser.

Over the years, the proportion of chemical fertiliser used on the fields has been progressively reduced, although at 200 tons per annum, still represents between 60 and 70 per cent of all fertiliser used. According to Xu Jianyu, this figure is currently being reduced largely because of the increasing costs of chemical fertiliser rather than because of its polluting qualities: farmers can earn more for the *jiti* and hence for themselves if they reduce their application of chemical fertilisers and use chicken excrement instead. Chemical pesticides continue to be used, however; indeed, the fields are sprayed with pesticides by aeroplane. Although a Dou Dian propaganda sheet of 1994 suggested that 'vegetable free from social effects of pollution is exported to Japan', Xu Jianyu, (Chief of Agriculture in Dou Dian) explained with a wry smile in 1995 that neither 'green' nor 'organic' vegetables were grown in the village. 'How would anyone know the difference?', he asked. However by 1997, Dou Dian had indeed begun to produce 'green' vegetables, using 'almost no' chemical fertiliser (less than 40 per cent of what was originally used), and only at initial seeding. Zheng Deyao (manager of the reception centre) explained that expansion of vegetable production had taken place on the one hand because there was a strong market for 'green' products in Beijing but on the other because grain production had become increasingly unprofitable because of increasingly high input costs and low market prices.

Animal husbandry still remains a key element of the ecological agricultural cycle as well as providing employment and output in its own right. Excrement is used as input to the biogas digesters or as fertiliser on the fields. 150,000 kilograms of grain and 350,000 kilograms of straw, much bought from outside the village is turned into animal products, providing markets in Beijing with 5,500 pigs, 5,000 beef cattle, 4,500 rabbits, 1.6m. eggs and 120,000 kilograms of fresh fish in 1995. However, the increasing costs of fodder is seriously denting the sector's profitability: while the beef cow farm is doing fine, selling meat to be eaten in the Great Hall of the People in Beijing, the profitability of the chicken farm and pig farm is now threatened as the costs of fodder go up.

160

8(4) 5. Political Economy in the 1980s and 1990s

8(4) 5.1 Political structures

As explained Dou Dian operates wholly as a *jiti* , its leadership, comprising 9 people, with *shuji* Zhang at the centre and including Xu Jianyu and the managers of the brick and garment factories, as well as a leader responsible for women's issues and family planning policy. This leading group overlaps with the Party committee, all are Party members and are subject to regular elections. It is the village leaders who, meeting on a day-to-day basis ultimately determine strategy: whether enterprises shall be established, expand, how much each enterprise should pay to the *jiti* and so on.

8(4) 5.2 Economic development

As suggested earlier, industrial enterprises quickly developed in Dou Dian in response to the need to find employment for the workers released from farm work following mechanisation. The brick works was the first to be set up, in 1979/80, followed closely by a building materials factory making prefabricated concrete parts. As these enterprises earned money for the *jiti*, various farm-related enterprises were further established including a slaughterhouse, an animal feed factory, a broiler and egg production factory and a dairy unit. And in 1983, a clothing factory, today the most profitable of all the enterprises in Dou Dian, was set up. By 1988, less than 10 years since the inception of the local economic reforms, 85 per cent of the workforce were engaged in enterprises, rather than working in the fields. By 1996, this figure, including animal husbandry, had increased to 97 per cent.

In the mid-1990s, Dou Dian has a thriving economy, based on a number of successful industrial and construction enterprises and animal husbandry which provide 95 per cent of the total income of the village. In 1993, the *jiti* turned itself into a company entitled The Dou Dian General Corporation of Agriculture, Animal Husbandry, Industry and Commerce with *shuji* Zhang Zhenliang as its first chairman. While farming contributed only a small fraction (i.e. 3 per cent) of total income in 1995, it remained profitable, unlike in many neighbouring villages where agriculture lost money. By 1997, however, it was losing money in Dou Dian as well and was being cross-subsidized from enterprise surpluses. Zheng Deyao argued that despite record harvests posted in recent years largely as a result of the increased productivity of organic over chemical fertiliser (the 1997 wheat crop had just been harvested before my visit in that year, producing 3.2m *jin* from 3170 *mu*, a little over 1000 *jin* per *mu*, comparing favourably with the 900 *jin* per *mu* average of the last few years) there were significant problems ahead for agriculture as losses mounted, largely the result of ever higher input costs, including costs of chemical fertiliser, and 'irrational' state pricing. By this he meant excessively low state prices, set in order to keep food prices down for the increasingly large number of urban residents ('to maintain political stability'), and this in 1997 (as in 1996) when state prices were *higher* than market prices: while the state price for wheat, depending on grade, was between 7.6 and 9.4

161

mao per *jin,* the market price was 6 *mao.*

Average income per head in Dou Dian was a little over 3,000 yuan in 1994, 3,800 yuan in 1995 and 4,200 yuan in 1996 (85 per cent from enterprises, 12 per cent from animal husbandry and 3 per cent from agriculture), making it the richest village in Fangshan District. The rise in the last few years represents a slowdown from the 1980s, however, largely the result, according to Zhen Deyao, of government restrictions on construction to reduce overheating in the economy.

While there are variations in wages paid by the *jiti* in different sectors of the economy - with the minority, a mere 98 workers, many women, still working in the fields earning the highest rewards - incomes are nonetheless relatively evenly distributed. Xu Jianyu explained that in neighbouring villages where the household responsibility system operated, there were individual households much richer than any in Dou Dian, but with the average household earning considerably less. The cooperative nature of the village organisation and the strong sense of social solidarity ensures that 'everyone gets rich together.' Meanwhile, there are no problems of unemployment and outward migration, unlike in many other neighbouring villages where people are moving (often illegally) to Beijing in search of work as agriculture ceases to be profitable. Indeed, Dou Dian now provides employment for large numbers of people whose residency *(hukou)* was originally outside the village. Nonetheless, according to Zhen Deyao in 1997, the population has fallen in recent years from 4,256 to 4,100, largely the result of the one-child policy and the fact that when students go to college in cities elsewhere, they don't come back.

In 1995 there were altogether 38 big and small enterprises run by the *jiti* in Dou Dian, ranging in size from the No. 9 Building Group Factory which employs over 1,000 workers (including construction teams, the building industry employs over 2,000 people) and the Yan Li Clothing Factory, which employed 900, to shops, restaurants and hairdressers with employees in single figures. The restaurant with the muslim-type dome, for example, is run by the *jiti* and functions as the official dining room in the village. The total annual output of the enterprises was approaching 100m. yuan. In 1997 the Yan Li Clothing Factory remains the most important in terms of revenue to the village. With its now 1,080 mostly young, female piece-workers, and 'advanced equipment' it makes anoraks, jackets and coats (altogether over 500,000 garments) which are exported, through an import-export company in Beijing, to 13 countries in all, including Germany, USA and Hong Kong. Meanwhile, the pharmaceutical goods factory makes a product called Da Guan mycini, which, it is claimed has 'curative effects on the venereal disease' and is known locally as 'efficaceous with one piece' (sic). Other important enterprises include the Brick and Tile Factory, the cement products factory and a factory producing enamel products and a carpet factory. The Muslim Meat Factory slaughters 10,000 beef cattle, 150,000 chickens, 6,000 sheep and goats while 150 tons of meat products can be stored in its freezers (Dou Dian, 1994).

Altogether 3,800 workers are employed by the *jiti* in Dou Dian, 60 per cent of which come from outside and live in dormitory accommodation. These are mostly young, single girls, from poor inland provinces, including Gansu. Wages average between 300-400 yuan a month in most of the enterprises; there is some degree of variation from one job to another depending on

162

the profitability of the enterprise: last year clothing factory workers earned between 450-500 yuan, similar amounts to those who work in the fields who are deliberately paid more than the average industrial wage, partly to increase the attractiveness of working there. Workers are allocated to jobs, as far as possible, on the basis of their choice, although ultimately managers of the enterprises have the right to hire and fire, and indeed to determine the wages of the workers. It was explained to me by Wang Encun that nowadays there were one or two private businesses *(geti hu)* in the form of restaurants, taxis and ice-cream salesmen but that 'they weren't encouraged' and that there were no privately owned factories. He suggested, however, that individuals in private business did work harder and sometimes earned more than the wages paid by the *jiti*. If a private household or individual wanted to set up a new business they had to apply for a licence from the village leadership; anything legal would be considered but anything polluting would not be granted a licence, he explained.

8(4) 5.3 'Social' development

As in most parts of the Chinese countryside, the material welfare of the villagers of Dou Dian has substantially increased in the last 15 years. Called by Xu and Simpson a 'former slum village' (1989, p.3), Dou Dian has built new houses and high rise apartments. In the 1980s, the housing drive concentrated on building single-storey traditional courtyard homes; more recently, because of the high building costs of so doing and the increasing shortage of land as the residential area has grown larger, the *jiti* from 1991 has concentrated on building high-rise blocks, using local labour and materials from local enterprises. The flats are popular and have been bought from the *jiti* at a cost of 700 yuan per square metre, providing income for further new build. In 1995, this new build was in progress and the prospective apartments already allocated. Most people still live in one-storey homes, but this is changing rapidly. Already the leadership has forbidden the building of new houses and Xu Jianyu predicts that within a few years most villagers will live in apartments.

Li Baozhong, a 77 year old great grandfather, and *shuji* from 1949 to 1952, proudly explained that in the 60s, his two parents, his wife, two children and himself lived in a mud house, without glass, unlit and unheated. His first brick house, built in the 1970s had only three rooms. Now he lives, as head of a four generation household, with his son and daughter-in-law, two grandsons and two great grandsons, in a large, traditional courtyard home, heated by coal burning stoves and supplied with biogas for cooking. While he himself is retired, his son is an accountant and his elder grandson a driver in Dou Dian. His living room is decorated with a statue of Mao Zedong and a picture of Jiang Zemin visiting his home in 1993 on the wall. (It was fairly clear that I was not the first visitor from outside to be invited to marvel at Li Baozhong's home and would probably not be the last. In that regard, he was not an 'ordinary' peasant. However it is clear that all Dou Dian residents are considerably better housed than before.)

1991 Dou Dian *cun* built a four-storey primary school to replace a dilapidated one-storey school: the fabric of the school is impressive with big, airy classrooms and is well-equipped: it has a language laboratory and has a range of PCs. A kindergarten is presently being built. Once the children leave primary school they go to Dou Dian *zhen* middle school. Dou Dian *cun* is also currently building a clinic; presently the clinic is run by Dou Dian *zhen*. There are currently no special facilities for old people although I was assured 'something will be done very soon.' According to Xu Jianyu, life expectancy is at least as high as anywhere else in China: 'there are many old people in Dou Dian.'

The one-child policy is strictly adhered to in Dou Dian. At the beginning there were great difficulties in imposing it, but now there are very few problems. One of members of the leading group is a woman responsible for the one-child policy; she visits women of child-bearing age and they are encouraged to attend the clinic on a regular basis. Were a woman to get pregnant with a second child, she would be expected to have an abortion.

8(4) 5.4 Environmental developments

Xu Jianyu is very clear that further progress in developing ecological agriculture is being and will continue to be made. When asked whether that development would change when the current (and aged) *shuji* retired from office, he was adamant that they would not, given the benefits that CEA had brought and despite the fact that, in recent years animal husbandry has become considerably less profitable. When asked what precisely the benefits of ecological agriculture had been, he emphasized higher cereal yields, with recent harvests providing more than 1 ton of grain per *mu*, lower input costs and biogas. To those benefits, Cheng Xu et al. (1992) add increases (to 1988) in biomass output, soil organic matter, soil nitrogen content and the rate of usage of crop residues for livestock feed, and despite an increase in the total industrialized energy use per hectare, a reduction in the required industrialized energy per kilogram of cereal grain produced.

Despite the commitment to ecological agriculture, however, and in recent years the commitment to the production of 'green' vegetables, Dou Dian is not an obvious model of environmental protection: its dirty streets, scruffy, if unpolluting industrial buildings and heavy traffic suggest rather the opposite.

8(4) 6. Conclusions

According to Wang Encun, in the 12 months up to July 1995, 10,000 visitors had come to Dou Dian from more than 100 countries. The numbers of visitors has warranted the establishment of a special unit to deal with visitors, with its own personnel, office, Jeep, and materials including videos (in Chinese; I was invited to watch them on my second visit). At Spring Festival in 1993, CP General Secretary Jiang Zemin was an honoured guest. During my first visit in 1995, several coach loads of Chinese visited the village and shortly after I left, the village played host to a group from the Philippines in China specifically to study CEA. When I asked why so many visitors came to Dou Dian, I was told that it was a model village and that that status was based on the high level of economic development, on the high level of mechanisation and on its *jiti* organisation. (Wang Encun did not mention CEA until I prompted him).

For whatever reason Dou Dian is deemed a 'model village', it is clear that it has not only made considerable economic advances but that some attention has been paid to reducing the extent of environmental pollution and degradation along the way. All those I spoke to were fulsome in their praise of the changes wrought in Dou Dian during the last 15 years. When in 1995 I asked Xu Guangkun, the leader of the pig factory and the father of 2 daughters, the elder currently in middle school, whether things were a lot better now, he replied with a huge smile (and unscripted, since there were no cadres about)

> Silly question. We earn more, eat more, get about more, are better housed, our houses are heated, our children well educated, with computers in the schools. What I spent in a year, my children spend in a day.

Explanations for the particular success of Dou Dian, both in economic and environmental terms, are not difficult to find. High on the list of any explanations, particularly of those inside the village, as so often in 'model' Chinese villages, is the quality of leadership. According to Dou Dian's propaganda the:

> rising of Dou Dian village is due to the wisdom of the leading group headed by Zhang Zhen-Liang (who)...since 1979 has won many awards including model worker and excellent Party member of the district, special grade model and excellent worker of party affairs; national model worker, national excellent leading cadre. Since 1982, he has been elected member of the standing committee of the 8th, 9th Peoples Congress of Beijing, member of the 6th Party Congress and representative of the 8th National Peoples Congress (Dou Dian 1994, p.3)

Though others are now taking part of the strain - Xu Guilan is *cunzhang* and is responsible for daily administrative tasks - the now 70 year old *shuji* Zhang Zhenliang has led the village, with a short break during the Cultural Revolution, for 40 years and he is still plainly held in great affection and respect by the villagers. Xu Guangkun (head of the pig factory) talked of his open-mindedness and his belief in science as critical factors, Xu Juanyu (head of agriculture) praised his genius. All those I talked to paid tribute to his contributions over the years.

In 1995, it was explained to me by the anthropologist and long-stay Chinese resident, Isobel Crook, who visited Dou Dian in 1986 after it had risen to prominence as a model village, that the reason for Zhang Zhenliang's success was because he had been a middle rather than a poor peasant before Liberation and that this meant, in so politicised a village, that he was for ever forced to stay in touch with the villagers and 'keep them sweet.'

Despite other qualities that Zhang Zhen-Liang possesses, what seems to have been the critical factor in Dou Dian's development was his interest in scientific developments in agriculture and his willingness to invite Cheng Xu and his team from the Beijing Academy of Agricultural and Forestry Sciences into the village to modernise agriculture along 'scientific' lines. Clearly, Dou Dian gained very considerably from the technical know-how and sponsorship of Cheng Xu's team (who became, for Cheng Xu & Simpson (1989), key 'change agents') over a 10 year period, first in terms of 'green revolution' techniques and the introduction of mechanisation and latterly in terms of developing an ecological agricultural system. While the village was still relatively poor in 1979 and while the Academy provided no injections of cash, the technological support over a long period was clearly decisive in Dou Dian's advances, both economically and environmentally. However, it was decisive not only in its own terms, but also in another crucial way.

When the HRS was introduced into the Chinese countryside in the early 1980s, Dou Dian refused to adopt it, primarily because the economic developments so recently made with the introduction of 'green revolution' techniques and mechanisation were popular and would have been put at risk by dividing up the land and property of the village amongst its households. As a result, Dou Dian remained *jiti* despite the pressure from the Beijing authorities which led every other village in the Fangshan district to adopt the reforms. And the importance of being *jiti* in the development of ecological agriculture in particular is not in doubt. It is plain that initially its adoption in the mid-1980s and the building of the first biogas digester were not popular amongst the villagers and that it only went ahead because of the determination of the leadership and their ability to steer it through. And clearly, Cheng and Simpson believe the form of social organisation in Dou Dian was vital:

> A crucial point is the recognition of Dou Dian's growth as a community-led project. At the project's inception, *the village was organised as a commune*, with essentially no free enterprise, not even individual marketing of garden products such as vegetables..... A critical factor has been the linking of farm services, such as providing tilling and harvesting equipment with a community-wide effort to expand income savings *for village use rather than just as an individual or personal activity* (1989, p.5, emphasis added).

Inside the village today, similar views are heard. When I asked Xu Jianyu in 1995 whether *jiti* organisation had been important, he answered with complete frankness 'obviously'. According to him, the communal organisation of the village has been crucial in three respects: it afforded the leadership the ability to overcome initial reservations, it has inspired a strong sense of social solidarity, and more pragmatically, enabled Dou Dian to afford the money to build the first biogas digester. And he was quite clear that the importance of *jiti* organisation is obvious to other villages as well and that lack of it has up to now effectively prohibited the development of CEA in them. According to him, it is very difficult for other villages to follow Dou Dian's example because they don't have enough communal funds to set up a biogas plant, they are likely to have insufficient organic fertiliser and even where they do, they don't have sufficient machinery to spread it, unlike in Dou Dian where two machines are dedicated to the task, bought from a factory in NE China subsequently closed. For Xu Jianyu in 1995, 'biogas must be set up communally. The *jiti* is vital. It is too difficult for other villages.'

In 1996, according to Zhang Zhenliang, *jiti* organisation was:

very important in Dou Dian's development and change to it was not possible. It allows planning and integration of industry, agriculture and animal husbandry.

In 1997, Zheng Deyao maintained the same line: as an employee and representative of Dou Dian yet a resident (and ex-*shuji*) of Dong Shi Yang, a nearby village which adopted the HRS in 1982, he was well placed to judge the relative merits of collective as opposed to divided land management. Although he still personally maintained a household level biogas pit, having encouraged, as *shuji*, the introduction of biogas in Dong Shi Yang (before it was introduced in Dou Dian), he was quite convinced that *jiti* organisation had been crucial in the development of biogas and CEA in Dou Dian. Only a dozen or so households still use biogas in Dong Shi Yang nowadays, he says, and then only his neighbours as a result of his advice and help.

Dou Dian has, of course, also gained from its proximity to Beijing in that it has always had ready markets for the products of its farms and enterprises close at hand and the commercial infrastructure of the municipality has allowed it to reach markets in other parts of China and overseas more easily than other locations would have done. The transport network is particularly favourable with good road and rail connections.

In summary, Dou Dian is a model village in many respects and CEA has been adopted pragmatically and gradually. Nonetheless, if Xu Juanyu and Zheng Deyao can be believed, it will continue to be developed and is in Dou Dian to stay, if only on the pragmatic grounds that the combination of machinery and organic fertiliser produces very high yields and because chemical fertiliser is increasingly expensive. It also has a name to live up to: its notoriety rests on many fronts nowadays, including the environmental front. However, whether the quality of the natural environment in other respects will be safe-guarded is another thing. In the twenty three paragraphs explaining Dou Dian's policy in encouraging foreign enterprise into its newly created industrial zone (Dou Dian, 1994), no mention is made of its environmental policy. And while CEA may take further hold in Dou Dian, on aesthetic grounds it is certainly difficult to sustain the idea of Dou Dian being in the lead in terms of environmental protection.

8(5) Teng Tou

8(5) 1. Introduction

Teng Tou is a small village of 795 persons in 275 households, situated in Fenghua County, in the low-lying north-east region of Zhejiang province, facing the East China Sea. The nearest city is Zhejiang's second largest, Ningbo, some 27 kms away, a city of 5.2m people, large enough to have its own airport, and with direct transport links from Shanghai by train (six hours) and ship (via the harbour facility at Zhenhai, ten hours). The only means of transport from Ningbo to Teng Tou is by automobile, but the new toll highway linking Ningbo and Fenghua,the county town, runs within a kilometre of Tengtou, so the village, while located in a wholly rural location, is currently well integrated into the rapidly developing economy of the region. The fast industrial and commercial development which has been characteristic of the eastern seaboard of China in the 1980s and 90s has made Zhejiang province one of the richest and fastest growing provinces in China with an annual GDP per head of 10,515 yuan in 1996 (China Statistical yearbook, 1998, p.65), ahead of its northern neighbour, Jiangsu, and surpassed only by the three municipalities of Beijing, Tianjin and Shanghai, and the province of Guangdong. The residential part of Teng Tou today is very compact, and in the summer, looks very pretty, with fruit trees, vines and flowers in evidence, with neat rows of modern two storey houses and fronted by a newly built village headquarters, which houses offices, meeting rooms and the guest house. It is surrounded by three small 'industrial' areas, 818 *mu* of arable land, 191 *mu* of orchards, 156 *mu* of forests and 66 *mu* of water surfaces (Teng Tou propaganda sheet, in Chinese, undated).

8(5) 2. Recent History

According to Fu Jialiang, the village leader and *shuji* for the past 43 years, the earliest records of Teng Tou date back 730 years. At Liberation, in common within the surrounding area and, indeed, most of rural China, Teng Tou was poverty-stricken. But unlike other neighbouring villages, there were no landlords in the village to be overthrown (i.e. they were absentee landlords) there were only three rich households and most of the then 417 inhabitants, all of whom had the family name Fu were middle or poor peasants. Land reform in Teng Tou was therefore a relatively gentle exercise, land being distributed to households on the basis of 2.6 *mu* per capita. Teng Tou remained poor, indeed had a reputation of being so poor throughout the 1950s that it was not an appropriate place for a man to seek a bride (Teng Tou propaganda sheet). In 1953 Fu Jialiang, at the age of 32 and a middle peasant, became *shuji*.

In 1958, at the start of the Great Leap Forward, Teng Tou was incorporated into the people's commune *(gongshe)* of Hong Qi, with some 30,000 people and one of the first communes in the county. In 1959 the commune was divided into brigades *(dadui)*. Teng Tou was so designated; the beginnings of a sense of self-reliance date back to this time. The experience of Teng Tou then was very similar to that in many revolutionary communes: 'everyone earned the same and did the same.' There was a communal dining hall from 1959 to 1962 and it was popular in the first instance because meals were free. But the place remained poor with frequent floods and 'natural' disasters.

In common with the experience of many communes, Hong Qi commune took a politically revolutionary line and when the call was made by Mao to make steel in backyard furnaces, some very small-scale iron foundries were established, trees were cut down to feed the furnaces and tools, farm and kitchen implements, even meal boxes were melted down to produce steel. According to long-term residents Fu Xingfan and Zhou Chanlin in 1997, people were 'crazy' but they had no choice since if they did not do so, they would be criticised and labelled anti-communist. The years of famine in the early 1960s, years in which bad harvests were compounded by lack of appropriate tools and implements, were years of 'serious distress' for Teng Tou, several households turned to begging and one 14 year old child starved to death. Though food was hard to come by, standards of housing were satisfactory however: even in 1960, there were one and two-storey houses which, though 'built of wood, ancient and not pretty', provided 'quite good' accommodation.

The experience of Teng Tou during the Cultural Revolution was no different from that of many other brigades: the Hong Qi commune took a strong political line in support of Mao and planting and economic development generally took second place to political activity. However, in 1966, at the start of the Cultural Revolution, Fu Jialiang made the decision that the brigade, on both economic *and* environmental grounds, should plant orange trees. This decision was contrary to the directives of the commune leadership who held that any economic activity other than the growing of grain and the raising of pigs represented actions along the 'capitalist road' *(ziben zhuyi dao lu)*. In other brigades, and indeed in Teng Tou before the decision to grow orange trees was made, politics was all important and 'people didn't pay attention to production.' They spent all their time reading Mao's books and put production aside. Some stopped production altogether. Thus Fu Jialiang was strongly criticised from outside the brigade for his decision to grow the trees, he was told by commune leaders and the leaders of neighbouring brigades to cut the orange trees down and was accused of being a 'capitalist tail' *(ziben zhuyi weiba)* and of using valuable land for growing non-grain products when they should be 'taking grain as the key link.' Criticisms of the brigade were particularly intense because Fenghua was the hometown of Chiang Kaishek. Despite the aggressive criticism from outside, however, Fu Jialiang refused to cut the trees down, arguing that they could do anything they liked to him, but they could not touch the trees. According to Luo Zhengyao (environmental advisor to Teng Tou), the *shuji* never thought about himself. And with the support of the brigade members who rallied around him, he carried the day.

Politics was not ignored in Teng Tou, and people read Mao's works, as in other brigades. However, 'people had to live' and in Teng Tou they did not stop producing. Indeed, the decision to grow orange and subsequently peach trees was merely one aspect of a range of actions which began in 1965 designed to promote the all-round economic development of the village. Even at that time, *shuji* Fu Jialiang recognised that the ecological environment was deteriorating: the quality of the soil was his main worry, rivers were increasingly impeded as sand dunes grew in size and number, ponds were inefficient and while excessive rain led to floods, the brigade additionally often suffered from drought. His decision to grow fruit trees, therefore, was not merely designed to increase the value of output but also to play a part in an overall strategy of soil conservation and water maintenance. From 1965 to 1978, with the 'enthusiastic' participation of the members of the brigade, the land was progressively flattened, fields reclaimed (over 200 cultivable sites by 1975), rivers controlled, irrigation ditches dug, ponds filled up and trees planted. In the 1970's double cropping, elementary crop rotation and mechanisation (Russian-made tractors) were introduced and every household built an individual biogas pit. By 1978, the immediate deterioration of the ecological environment had been arrested and the economy was in a sound, 'middling' state, with grain output per *mu* over 500kg. rice annually.

Moreover the all-round development of the brigade was reinforced by the establishment of factories in the early 1970s, albeit clandestinely: the first factory made building materials, it kept machines in cattle sheds and ensured all its workers were senior cadres and party members who kept its operations quiet. According to Fu Xingfan and Zhou Chanlin, this was a common practice because the Fenghua County Party was very strict and it was important to appear to follow the party line. However, as industries expanded to include an industrial diamond factory and a clothing factory, what was going on became public knowledge and the brigade suffered alot of criticism from outside. Before the end of the Cultural Revolution *shuji* Fu Jialiang reportedly went to a meeting in Fenghua, was refused entrance and had his luggage thrown out. Nonetheless the criticism did not stop the brigade developing enterprises although the diamonds factory was temporarily shut down.

8(5) 3. Land Reform

Although Teng Tou was renamed a village *(cun)* in common with all the neighbouring brigades of Hong Qi commune in 1982, its fields were distributed to individual households with the introduction of the HRS *(chengbao daohu)* as early as 1979. The initial division of the land - the forests and the ditches remained in the hands of the brigade and were managed by the *jiti* - was popular amongst the brigade members who enthusiastically embraced specialised raising and planting. The distributed plots, however, were small, in the region of 10-20 *mu* depending on the size of the household. Farmers soon found that they now had to buy their own tools and, if they had insufficient organic matter, their own fertiliser. Since plots were so small, mechanised methods of cultivation were inoperable and farmers soon found 'that they had to put alot in but did not get alot back.' And as early as 1982, difficulties appeared. In that year, as working in enterprises became an increasingly popular option, some households allowed their contracted land to be recollectivised under the *jiti* and were left with only eight *fen* per capita to use for their own devices.

As the all-round economy of the village developed throughout the 1980s farmers found it more and more profitable to desert their land in favour of taking up employment in the various industrial, animal-rearing and transport enterprises set up and managed by the *jiti*. As they did so, they progressively gave back their land to the *jiti* which in turn expanded its role in land management. The giving up of the land was a gradual process but a consistent one. Indeed, households not only gave up their contract land but their private land too. By 1994, the process was complete: all agricultural production, in common with almost all other forms of commercial production, was run by the *jiti*, all land had been recollectivized (with no apparent resistance, explained by villagers' confidence in the *jiti* and its *shuji*) and was farmed by the *jiti* which was responsible for fulfilling the state quota. 'Ownership' had gone full circle in just 15 years. By 1997 the land had been parcelled out to just five 'big households' *(dahu)* to farm on behalf of the *jiti*.

Shuji Fu Jialiang admits that in the late 1970s, he was opposed to the policy of dividing up the land in Teng Tou as was required by the policy of *chengbao daohu* but was forced to go along with it as a result of pressure from above, 'you couldn't fight against the policy' he says. His opposition to the policy rested largely on the grounds that by 1979, Teng Tou was a reasonably rich area based on mechanisation which would (and did) become redundant once the land was divided up. He suggests that in many such rich areas, working on the land became less attractive and led to farmers not wanting to work there. Asked whether the policy was therefore a mistake from the start, Fu Jialiang suggests that in many poorer villages it was not a mistake, as villagers for the first time enjoyed incentives to produce more, but that it was a mistake to apply it to rich villages, that it was a mistake to apply *chengbao daohu* in a blanket fashion without regard to circumstances. Different policies should have been applied to different villages. Asked whether the villagers in any way resisted the gradual encroachment of the *jiti* in the 1980s, he replied in the negative, suggesting that every year as the village developed they could see the benefits and that the increasing role of the *jiti* was popular as a result. Fu Xingfan, Zhou Chanlin and Luo Zhengyao all agreed with this analysis, the latter arguing that the *shuji* was 'very democratic' and that many village meetings were held of an educational, political and spiritual kind. They all accepted that the local Party branch had played a very significant role in the village from the start and continues to do so.

8(5) 4. Ecological Agriculture

Though it is clear that economic developments in Teng Tou from 1965 onwards were informed by the need to create and maintain a strong ecological environment, it is argued by Fu Jialiang and Luo Zhengyao, (loyal to the official line) that the key moment in the acceleration of development, both economically and environmentally, was the 3rd Plenum of the Central Party Committee in December 1978. Ecological agriculture, as it is presently conceived, was gradually adopted from 1979 onwards, even though this was done in a largely unconscious way, the official term as yet not having been coined. Inspired more by his concern for soil quality than by reading about the subject, the main instigator of these developments was, again, *shuji* Fu Jialiang. According to Luo Zhengyao, 'the *shuji* was doing ecological agriculture without knowing it.' The village had no outside assistance.

The most significant development which took place in the period from 1978 to 1981 was the digging of a fully concreted ditch around the residential part of the village, 830m. long, 6m. wide and 2m. deep with 6 km. of underground irrigation pipes linking it to taps centrally placed in the fields. 4,000 orange trees were planted on the banks of the ditch and where at intervals concrete frames two metres high had been constructed, grape vines were introduced. Altogether 837 grape vines were planted. Meanwhile, building on ecological achievements of the 1970s, the brigade progressively replaced chemical fertiliser with organic fertiliser in the form of river mud, barnyard manure and stalks to the extent that 15 years after 1980, 120 tonnes of organic fertiliser had been applied to each hectare of land. At the same time, experiments took place to develop biological alternatives to pesticides, introducing bull-frogs, spiders and snakes. New crop rotations, such as rice-rice-rape or rice-rice-green manure were introduced and multi-layer systems of cultivation and/or raising developed. In the hills to the immediate north of the village, for example, conifer afforestation took place on hilltops while bamboo trees were planted on the slopes and fruit trees at the bottom. Closer to the village, ponds were dug and fruit trees planted at their edge, ducks raised on the surface while different kinds of fish swam in the water at different levels. These processes were referred to as 'vertical' cultivation.

Throughout the 1980s, there was continued emphasis put upon the recycling of wastes, as less chemical fertilisers were applied to the fields and pesticide application was reduced. Animal rearing developed only gradually. Fish, shrimps and ducks were introduced in 1979 and in the early 1980s the *jiti* took over some specialised pig-rearing. However major developments in animal husbandry did not take place until the end of the decade. In 1989, the village established a new pig farm, with 1250 fine-breed pigs, growing to 2000 in 1990, and in the same year chickens and cows were introduced for the first time. By 1990, animal rearing produced an output value of 283,000 yuan, 55.2 per cent of gross agricultural output.

With the establishment of the pig and chicken farms came the decision to build a community level biogas digester. While individual household biogas pits had been constructed in the 1970s, they had not been very successful largely because they were difficult to run without regular supplies of appropriate organic matter and because farmers found technical problems difficult to overcome. Thus, in 1992, 650,000 yuan was spent by the *jiti* on building a community level digester with a capacity of 500 cu.m., using the night soil from the pigs as input to produce gas to households for cooking and lighting, solid slurry to be returned to the fields as fertiliser and liquid waste to be piped into the underground irrigation system. Teng Tou had effectively adopted an archetypal ecological agricultural system.

8(5) 5. Political Economy in the 1980s and 1990s

8(5) 5.1 Political structures

The village itself is variously a social unit and a commercial unit: the term *jiti* is used to refer to the collective nature of the social unit, yet commercially the village trades under the title of Teng Tou Industrial and Commercial Enterprise Company and the leaders form its board. In 1997 *lao shuji* Fu Jialiang is chairman, but deputy *shuji* Fu Xiping is President *(zhong)* of the Board. (It is not entirely clear the division of responsibilities between these two posts, but Fu Jialiang at 73 is probably less active and his role more ceremonial than that of his 45 year-old colleague). Fu Aijun is village leader *(cunzhang)* at the same time as managing the property company owned by the village. The leading group of Teng Tou today comprises a further five leaders, one of whom is the manager of the clothing factory, Teng Tou's largest enterprise. All are members of the Communist Party, as are the secondary level leaders both within the village and in the factories. It appears the party maintains a very strong grip on affairs and as in other Chinese villages, there is a strong overlap between the political and economic functions of the leaders.

8(5) 5.2 Economic developments

As early as the 1970s, a range of sideline industries were set up, albeit clandestinely by the village and run cooperatively by the *jiti*, including the building materials factory, an industrial diamonds factory and a clothing factory. A photo-processing factory was established in 1982, and in the next 3 years there followed a cardboard box factory, a leather goods factory, a down garments factory, a flooring products factory, making use, as far as possible, of local products and resources and a transport enterprise. As industry developed, three industrial sites were designated, A,B and C, the first two of which were (and still are) located in the centre of the village, cheek-to-jowl with the residential area, the third at a distance, 'in the hills.' The latter location was chosen by the *shuji* on the grounds that building on arable land close to the village was a waste of good land.

Whilst the CEA system remains an important element of the village economy, the main source of its income is now firmly its industrial enterprises. In 1997 there were 16 enterprises run by the *jiti* on the three industrial sites, employing altogether 1,300 employees. Seven of these enterprises are now joint ventures with foreign companies. Many of these employees are women bussed in from neighbouring villages, the villagers of Teng Tou being largely in managerial and supervisory positions. The factory workers from outside Teng Tou are paid 400 yuan a month, while the average emolument of the Teng Tou villagers engaged in industrial work is about 700 yuan a month. Experimentation is taking place with regard to share ownership: some of the managers, according to Fu Jialiang, are buying shares in their factories and enjoying a cut of any profits.

By far the largest and most important of these enterprises is the clothing factory originally set up in 1985, now a Sino-US joint venture employing 600 workers, and making suits which won a gold medal in the Leipzig Fair in 1990 and which are regarded as some of the best products in China. Apart from suits, coats and knitwear, it makes the uniforms worn by the guards in the Great Hall of the People in Beijing. In 1994,with its products being exported to 14 countries the total value of its output was 120m. yuan, making a profit of 8.51m yuan. Another important Sino-US joint venture, set in a prize-winning 'garden factory' is the artificial diamonds enterprise (Diamond Tools Co.Ltd), employing 30 workers and producing an output value of 8m yuan and a profit of 1.5m. yuan in 1994. The latter figure rose to 2m. in 1996. Other important joint venture enterprises, with either Hong Kong, Japanese or South Korean partners, produce chemicals, down products, leather goods and cement tiles. Other enterprises include the photo-processing factory, the cardboard box factory and the grain-processing factory. Foreign investment in Teng Tou began in 1990 with the US involvement in the clothing factory after it had been nominated, on the recommendation of the province, as a National Model Village for its achievement in 'spiritual civilisation' by the Ministry of Urban & Rural Construction. More recently, after the village won the UNEP 'Global 500' award in 1993, it attracted a new wave of interest from foreign investors. Since 1993, Teng Tou has attracted 10m.yuan in foreign capital into its seven joint-venture enterprises.

In 1994 the total value of output in Teng Tou was 204m yuan, almost double the gross figure for 1993, the level of profit was 17.88m yuan (up 71 per cent on 1993) and the average income per head 5,026 yuan, (up 22.5 per cent), 85 per cent of the value of output is attributed to industrial enterprises and 60 per cent to the value of the output of the clothing factory alone. Income per head in Tou Teng is thus about 80 per cent higher than the average in Zhejiang province (an already rich province), making it the richest village in the Xiao Wang Miao township *(xiang)*. In 1996, Teng Tou boasted an income per head of 7,016 yuan, making it by far the richest eco-village researched. It is claimed that its economic power is on a par with a normal *xiang*.

In 1997, about 10 per cent of the households were involved in 'private' *(geti hu)* activities viz. small shops, hairdressers, restaurants, taxis. In 1996, a branch of the Bank of China opened in Teng Tou, surely a milestone along the road of 'economic development' for *any* Chinese village. Average accumulated savings in Teng Tou are 20,000 yuan per household.

One of the maxims stressed by the village leaders is 'without agriculture, no stability, without industry, no riches' *(wu neng bu wen, wu gong bu fu.)*. Although there were years in the 1980s when agriculture in the village was supported by transfers from industry, nowadays this is no longer the case. Indeed, in 1994, the value of agricultural output and animal husbandry was 4.5 yuan yielding a profit to the *jiti* of 570,000 yuan. There are in 1997 610 *mu* of fields, 400 *mu* planted to rice, 50 to 60 *mu* of orchards, the rest vegetables. Animal husbandry, in particular, pig raising has gone from strength to strength. In 1997, the village was raising 5,000 baby pigs, i.e. of the weight of 95-100 kilos, all for the Hong Kong market. Two truckloads leave Teng Tou for Hong Kong every week and the plans are to double this production by 1998, a process which will involve an expansion of the biogas digester to cater for the increase in excrement (already, there are three tons of pig manure produced every day).

Villagers all earn roughly equal salaries of about 700 yuan from working in enterprises, although there is a little disparity, related to the profitability of the enterprise: salaries are highest in the clothing factory. Managers earn 'a little more' than workers, but not much more. Salaries paid to the agricultural workers are the highest of all. Villagers pay no taxes, although they pay 20-30 yuan a month charges for electricity, water and biogas. Retired workers receive a pension from the *jiti*.

Teng Tou has lost its earlier reputation: it is now a village where outsiders actively seek future marriage partners. House construction also took place in the 1980s: the main housing drive began as early as 1979 and over the next 6 years, 3.1m yuan was invested in new houses. Altogether 252 new houses were built all in the nucleus of the village, one for each household, each house generously proportioned with 116 square metres of floor space, allowing 39 square metres per person. Houses are all two-storey and provide living space far in excess of the average in rural areas (and even more so than in urban areas). In 1991 plans to rebuild the houses were shelved on the advice of that year's most honoured visitor, CP General secretary Jiang Zemin, who felt the standard of housing was already good enough. In 1996 the gabled roofs were be replaced by flat roofs to provide more space for domestic cultivation but this had not been done by summer 1997.

With regard to other public buildings, a new kindergarten and primary school have been built in the last 5 years. In 1992, the village built its main administrative headquarters, housing offices, the village canteen and a well-equipped hotel. Along with all the other key leaders *Shuji* Fu Jialiang has his office in the building. [While I was told that the visiting UN team had been very impressed by the frugality of the *shuji's* living accommodation; his rather grand office also with en-suite facilities, could not be described as frugal.] The building also provides an activity room for old people and a laundry. By 1996, a village museum was completed, as was a new 'cultural centre' which includes a dance hall and karaoke room. Villagers can enjoy services normally associated with much larger towns.

The leadership suggests that while it was difficult to enforce the one-child birth control policy in the early years of its existence, it has been largely adhered to in recent years, primarily because, as the village has got richer and the nature of the work has changed, children are increasingly seen as an economic burden rather than as an asset. Nonetheless it is clear there is still considerable coercion involved: a family is stopped from having a second child if the first is a boy, and must wait for seven years for another if the first is a girl.

Teng Tou has continued the all-round development of its economy in the 1990s and is now a village increasingly relying for its rapidly rising income levels on industrial enterprises. However it was as an exemplar of a village successfully developing CEA that Teng Tou won the UNEP 'Global 500' Award in 1993. The process of development began 30 years ago in 1965, with ecological agriculture at first an unconscious partner and subsequently, through the 1980s, playing a more conscious and explicit role. Reportedly, Teng Tou developed CEA largely on its own, with little help from outside.

Today, with the development of animal husbandry and the construction of the community level biogas digester, the village has a fully integrated, archetypal ecological agricultural system. The fields producing grain (now almost exclusively rice), and since 1994 completely reorganised on *jiti* lines, are today cultivated by just five households, chosen by the village leadership. According to Fu Jialiang, the work is 'not hard' given the high level of mechanisation: there are over 40 farm machines of various kinds and the crops are harvested wholly mechanically. Meanwhile, water for the crops is immediately available from taps from the underground irrigation system. Those villagers engaged in agricultural production are, like every worker in Teng Tou, paid a salary (in 1995 of 750 yuan a month, 100 yuan a month more than the average paid to Teng Tou villagers working in industrial enterprises) and they produce the rice entirely for procurement by the national government, not directly for the villagers to eat. If anyone in Teng Tou wants rice, they buy it, just as urban residents do. According to the leadership, the fields are nowadays mostly fertilised organically, through the planting of green manure, the partial return of the stalks to the fields and from biogas slurry. This, combined with careful use of crop rotation, including rotations involving green manure, has allowed steady increases in yields and significantly improved soil quality. Although there is negligible use of chemical fertilisers, modern pesticides are still used in Teng Tou given the lack of complete success to date in developing biological alternatives.

Apart from crop rotations, 'multi-level' ecological agriculture remains widely practised in the growing of fruits and vegetables: orange trees and plum trees and orange trees and vegetables are grown in the same fields. In one field, pear trees, watermelons, beans and peanuts were all in evidence. Other 'multi-level' processes include the widespread planting of fruit trees on the banks of the ditches and ponds, with ducks swimming on the waters' surface and shrimps and fish being raised at different levels in the ponds. In this cycle, the night soil of the ducks feeds the fish while the accumulating mud at the the bottom of the ponds is used to fertilise the fruit trees. Fruit trees, in particular oranges, peaches, apricots and vines are everywhere in evidence, on the roadsides, the banks of ditches and in the front of the houses. Flowers are also grown as are a huge variety of ornamental cactuses, produced primarily for export to Hong Kong Meanwhile multi-level processes also operate in the hills on the village border, with the growing of fir, bamboo and fruit trees at different altitudes, while indoors, chickens are raised on the upper floors of the animal houses with pigs, feeding off the night soil of the chickens, being raised on the lower floors. Pigs are no longer kept by individual households. If the villagers want to eat meat, they buy it in the shops.

As in all arguably complete Chinese ecological agricultural systems, biogas is at the hub. The new biogas digester, built in 1992 and, with a capacity of 500 cu.ms., the biggest biogas digester in Zhejiang province, has a daily output of 1,000 cu.ms., and provides all households and the village dining hall with 'clean' gas for cooking. The main input for the digester is the night soil of the pigs, three tons daily! Altogether 9 people work full-time at the digester and they claim to have overcome all the technical problems that have arisen to date. Every household pays 13 yuan a month for receiving biogas; before, households cooked primarily by burning stalks which were collected free. Meanwhile, the biogas provides solid slurry which goes straight to the fields as solid fertiliser while the liquid slurry is piped into the underground irrigation system. Given the availability of organic fertiliser, the village has begun to grow 'green' vegetables and in 1997 established a 'Green Food' Company. Not only is the biogas clean, but it is efficient. I was reliably informed that while it took nine minutes to boil a kettle using an ordinary gas canister, it took only seven minutes with biogas!

The biogas operatives and the village leadership are very bullish about it. The latter see the biogas digester as being a good advertisement for the village and they run training classes on biogas for people from outside the village. Zhou Huada, the leader responsible for Public Relations and a member of the village environmental protection committee, whilst delighted with the success of the biogas digester suggests, however, that biogas is not a sine qua non of ecological agricultural development. He is aware that the initial investment is very expensive, that technical problems recur and that many villages will incur great difficulties in finding inputs of appropriate biodegradable materials on a sufficient scale (most notably animal night soil), particular since the adoption of the HRS. The fact that Teng Tou is run entirely on *jiti* lines and has built itself a large pig farm is, he claims, a great advantage in meeting these input requirements. However, he suggests that the village has spent a lot of money on the biogas digester, is continuing to spend money given the technical problems and can only do so because the village is rich.

The village won the UNEP 'Global 500' award in 1993 and the *lao shuji* Fu Jialiang collected the honour at the Great Hall of the People in Beijing from UNEP Executive Director Elizabeth Dowdeswell on World Environment Day, June 5, 1993. Despite the obvious importance of industry to the economy of Teng Tou, the current leaders insist that ecological agriculture will remain at its core and that there is no likelihood of any deviation from that, citing the recent huge investment in the biogas digester as evidence of their commitment. To win the UNEP 'Global 500' award is seen as a great honour. Moreover, they insist that polluting industry will continue to be prohibited, despite their ambitions for further industrialisation. It was explained that in 1994 an Australian company which had plans to invest $1.4m in a joint venture project making silk materials was refused because it would have involved the emission of 500 tons of polluted water a day. And it is claimed that between 1994 and 1997, 15 factories were refused permission to set up because of their potential for pollution. They are aware that, while this policy may involve prohibiting certain income-producing projects in Teng Tou, much of its recent industrial investment from abroad has occurred *because* of its fame as a model ecological agricultural village, and as a result they are concerned to do nothing to compromise that.

8(5) 6. Conclusions

All the Chinese experts I interviewed on the subject of Teng Tou (in NEPA, Beijing and Nanjing) held very good opinions of its recent development and from direct observation in the village, it is a thriving economic unit paying considerable attention to the quality of the natural environment. The social infrastructure boasts comfortable housing, schools and an array of public buildings for the benefit of villagers of all ages. Its ecological agricultural system, with the community-level biogas digester at its core, is archetypal, and has been rewarded with provincial, national and international interest and praise. In 1996, it successfully hosted a conference for all past UNEP Global 500 winners, in 1997 it won, alongside only three other villages (in Jilin, Hainan and Shaanxi) the title of 'model socialist village' from NEPA, meriting the award on the basis of its good spiritual civilisation, agricultural development, CEA development, good party organisation and equality in the distribution of its income. According to Fu Jialiang, the evidence of the 'good spiritual civilisation' is to be seen in the high moral standards, polite behaviour and in the absence of gambling, stealing, superstition and fighting. And others are learning from its example: it receives many visitors -it claims to have had 26,000 visitors in 1996 alone. In 1991 Jiang Zemin was its most famous visitor, in 1993 it revived a delegation from the UNEP which included Elizabeth Dowdeswell.

In order to explain this successful development, there are a number of factors to be discounted. It was not a particularly rich village at the start of the 1980s and its progress in developing ecological agriculture, animal husbandry, industry and social infrastructure was gradual, depending neither on a rich seam of accumulated savings from past decades nor on a sudden influx of financial or technical help from outside. It was not a village adopted by an experimental environmental research institute or environmental protection bureau and it claims to have received no outside help nor to be currently dependent on any outside agencies. Moreover, while it is located in an economically rapidly developing region of China and is currently able to benefit from the infrastructural improvements taking place, its geographical location is not especially favourable: good road connections to Ningbo, the largest city in the neighbourhood have only just been constructed while Ningbo itself is less accessible from major commercial centres like Shanghai than many other cities in the region.

The most striking feature of the political economy of Teng Tou is the strength of the *jiti* and the power of its leadership. The land reforms of the late 1970s have been completely overturned and since 1994 the village has operated a wholly collectivised agricultural system. This has allowed mechanisation to occur and economies of scale to be realised. Only five households farm the land full-time; everyone else is in gainful employment in profitable side-line industries.

Meanwhile, it is the collective system which has allowed the successful adoption and management of the community-level biogas digester. The problem of employing farmers released from the land has been solved by industrial development. Indeed, such is the scale of industrial employment that 1,200 workers, mostly young females are employed from outside. These outsiders, most of whom live in dormitories, have their temporary *hu kou* in the village but do not enjoy the normal benefits of residence (such as schooling etc.) nor the same rates of pay as the villagers. Today, even if it is ecological agriculture which has brought Teng Tou fame, it is successful rural industry which underwrites it.

The village enterprises are all owned and managed by the collective leadership of the Teng Tou Industrial and Commercial Enterprise Company. The Party remains strong in Teng Tou. According to the Luo Zhengyao and Zhou Huada, this is no accident, indeed, the success of the village is attributed to the strength of the *jiti* and the quality of its leadership, particularly that of *lao shuji* Fu Jialiang, leader since 1952, who 'through his energy, ideas and farsightedness has always been able to persuade others to work together in the best interests of everyone in the village.' Zhou Huada explained that he had been to many villages where the leaders had got rich but paid no attention to ecologically sound agriculture. Fu Jialiang has stayed close to the villagers and lives amongst them, in the same conditions.

Fu Jialiang, in common with some other village leaders mentioned in this work took an independent and progressive line during the Cultural Revolution to the benefit of the village and it is clear from direct observation of his interaction both with other leaders, visitors and villagers that he commands almost unique respect and affection from everyone who knows him today. It is difficult to come to any other conclusion than that his imaginative, responsible and trusted leadership has been of primary significance for the environmental and economic development of Teng Tou. According to Zhou Huada, getting villagers to work together with the leadership is not always easy and he argues that difficulties have been overcome in Teng Tou largely as a result of the high priority given to environmental education and training, both in the school and the factory. All the villagers are expressly educated in the benefits of CEA, with Luo Zhengyao active in this regard.

It is thus argued that the reconstitution of the *jiti* and the development of the *whole* village, rather than the aggrandisement of a few individuals, has been of critical importance in explaining the recent successes in Teng Tou. According to its present leaders, a strong *jiti* is the *only* way ahead for other Chinese villages.

8(6) Qian Wei

8(6) 1. Introduction

Qian Wei is a small, modern village of 759 inhabitants in 279 households at the northern end of Chongming Dao, on China's 3rd largest island (after Taiwan and Hainan), lying at the mouth of the River Yangtse ((*Chang Jiang)* and within the municipality of Shanghai. Chongming Dao is flat and very low-lying, much of the land to the north (including Qian Wei) having been recently reclaimed from the sea. The island is reached by boat from Shanghai, the sea journey from Baoshan, one of the ferry ports north of Shanghai, to Chongming town taking 40 minutes by hydrofoil and 1 1/2 hours by passenger ferry. Buses run the 30 kilometres from Chongming town to Qian Wei although aggressively competitive taxis will do the job more quickly and comfortably. Despite being a small village some distance from Chongming town, the name Qian Wei is well-known there, having been designated in 1994 Chongming Dao's 'first village' and in recent years being the recipient of many visitors. Though the economy in Chongming Dao is diversifying into light industry and tourism (and being led in so doing partly by Qian Wei), it is still primarily agricultural in orientation, with rice, wheat, vegetables and fruits as the main crops. In 1994, Chongming Dao was designated by NEPA as one of the 50 'ecological agricultural counties' and Qian Wei itself was chosen as a possible recipient by NEPA of the UNEP 'Global 500' award in 1996 for its development of ecological agriculture.

Qian Wei's total land area is 3271 *mu*, although fishponds account for a further 550 *mu*. It is a village which looks like few others in China, in that its main residential and industrial area is bounded by walls and hedges and is entered through recently constructed village gates staffed by officials. Entering the village necessitates showing the gatekeepers documentary proof of identity and invitation: it is more like entering a high-profile work unit *(danwei)* than a village. This is not surprising in that the economy of the village, despite its notoriety as a model for *ecological agricultural* development, is highly geared up to a range of industrial and commercial activities collectively *(jiti)* owned and managed by the village as a unit. The nucleus of the village is very compact, with the factories, the public buildings e.g. the administrative block, the restaurant, the two guest houses and the residential housing all within a few metres of each other. This central area, mostly of recent (1980s and 90s) construction, is surrounded by the fields and ponds.

8(6) 2. Recent History

Qian Wei is a *product* of the Cultural Revolution. Until the end of the 1960s, the village did not exist. Then, in 1969, a work team of 70 'volunteers' from the Da Xing commune was organised to come to this part of the island, which was in the geographical boundaries of Da Xing, to engage in a process of land reclamation. Being low lying and close to the sea, the area was crossed by a series of small rivers and was flooded for long periods. The land was built up and converted into fields and altogether 424 mu. was reclaimed for planting in that year. At the same time rudimentary (mud and straw houses) houses were built and the brigade *(dadui)* was founded with the 70 volunteers as its first residents. Its original name was Sheng Li *dadui*. Two of these original brigade members were Xu Weiguo, who in 1973, aged 21, became village leader and *shuji*, and who remained in those positions until 1993, and Lu Shanxi, who is today in charge of the community level biogas digester in Qian Wei. According to the latter, working and living conditions were very hard in the first few years and income was very low. Through the 1970s some brigade members were sent outside the brigade to do the 'dirtiest' jobs purely to earn extra money. But those left working the fields also had a hard time and average incomes, working inside or outside the village, were no more than 25 yuan a month.

8(6) 3. Land Reform

Sheng Li *dadui,* despite its ambitious leadership, remained poor throughout the 1970s and when, in 1982, land reform and the HRS were ushered in, division of the land was eagerly undertaken. As Lu Shanxi explained, land division was very popular because 'if you worked hard, you earned more and once you had paid taxes everything belonged to you.' However, the enthusiasm for cultivating the very small plots (between 1-2 *mu* per head) did not last long given the high costs of production, in particular the cost of fertiliser, and the superior and increasing attractions of working in the factories which were being established in the early 1980s (see below). Indeed, according to Xu Weiguo, grain production actually went *down*. Within one year of the introduction of *chengbao daohu,* the first plots of divided land were transferred to specialised households as farmers found they could earn more in the factories. Soon the exodus from the land became a generalised phenomenon and given the need for factory workers, Xu Weiguo became convinced of the need to reunite the fields under *jiti* control. By the mid-1980s, this had been achieved without opposition either from within or from the authorities, and workers were recruited from outside the village not only to work in the factories but to cultivate the land. Food was grown collectively and provided to individual households, as now, at minimal cost. Sheng Li *cun* was the only village in Da Xing township to revert to *jiti* agriculture at that time and Qian Wei remains the only *jiti* village on the island of Chongming Dao to this day.

Sheng Li *cun,* as a result of developments described below, had by 1993 achieved sufficient economic strength that two smaller neighbouring villages, Bao Zhen and Da Tong, were incorporated within it. At the same time its name was changed to Qian Wei.

8(6) 4. Ecological Agriculture

Ecological agriculture was not consciously adopted until the early 1990s and postdated industrialization in Qian Wei, the latter arriving very early. It was the very harshness of life

181

which predisposed Xu Weiguo to find a new way for everyone to get better off. From the start, it was his idea that people should get better off *together,* and he realised that it would not be possible to get rich on the basis purely of agriculture. In common with the conventional sentiment ubiquitously expressed, it was the outcome of 3rd Plenum of the 11th Communist Party Central Committee in December 1978 which paved the way for a new beginning. In 1979, at Xu Weiguo's instigation, the brigade set up a chicken farm and four small food processing factories. However, they were not successful: the farmers were simply unable to organise and manage good factories, lacking as they did, the necessary managerial and marketing techniques, experience, ideas and materials. It was this experience which encouraged the leadership to search for an enterprise outside the village which did have these things and with whom they could set up a joint-venture. But it was initially not easy and though they tried many times, costing the village over 70,000 yuan from their accumulation fund in putative ventures, they had no success until 1982 when they teamed up with another village, Chang Zheng, which, already having 13 factories, brought not only the skills and techniques lacking in Sheng Li (Qian Wei), but 185,000 yuan as initial capital investment to set up a factory making shoe polish there. In the first full year of its operation, it earned 100,000 yuan sales revenue. By 1986, its turnover was 832,500 yuan and it had become a major source of non-farm employment, employing up to 100 workers at various times of the year.

Shu ji Xu Weiguo decided that it was vital to expand the industrial activity of the village. Initially the leadership considered establishing a factory to make toothpaste and to do so entirely on its own but there were difficulties in getting hold of the necessary materials and as a result plumped for a joint venture arrangement with an enterprise from Shanghai already making toothpaste in the city. As a result, in 1986 a new factory making toothpaste was built in Sheng Li, soon to employ 200 workers, mostly local people but including a few experts from outside, including from Shanghai. The factory was immediately successful. In the first year of its operation, its output value was 1.1m yuan, by 1991 this figure had increased to 15m. yuan. The success of the toothpaste factory at that time is now put down to a combination of three things: management skills from Chang Zheng, technical skills from Shanghai and the hard work and enthusiasm of the local villagers. The leadership stressed the importance of inviting experts from outside, aware that industrial employment and high industrial incomes depended not only on the quantity of output but also on the *quality* of workmanship. They were aware that the farmers were essentially materialist but that if the village was to prosper, material incentives were not enough and farmers needed organisation and training from outside. The strategy appeared to work: the village accumulation fund which had stood at 35,000 yuan in 1981 had reached 65m yuan by 1994.

However, the road to that latter figure was not a smooth one. In 1992 disaster struck when, after a period of official reining back of aggregate demand in response to the general overheating of the economy in the late 1980s, the toothpaste factory found great difficulty in selling its output and was forced to shut down production for 3 months. Xu Weiguo and the leadership were determined that the economy of the village should not go down with it and as a result the key decision to diversify was made. They decided to take a high science-based, biotechnological route and began to work in cooperation with local scientific institutes, including the Life Science

Institute of Fudan University in Shanghai to develop new industries. Two key developments resulted in 1993: first, in conjunction with Fudan University, the toothpaste factory was relaunched producing a new kind of toothpaste (FE), which, according to its own packaging 'is formulated on principles of life sciences, of super quality with an active ingredient, FE Biosin, produced by advanced biotechnologies.' Secondly, a factory producing *ru suan*, or lactic acid (DL), a colourless crystalline organic acid produced by anaerobic metabolism of glucose and designed to be a nutrient and preservant in soft drinks, was established at a total cost of 4m yuan. The development of other biotechological processes quickly followed, including the setting up of a medical institute in the village by twelve experts from a Shanghai hospital, with the purpose of developing nutrients to increase life expectancy. By 1995, the output of the DL factory had still to come on stream; the FE toothpaste factory was immediately successful, however: in 1994, it was designated a 'model product' in Shanghai and won a prize from the National Science Commission.

The successful establishment of biotechnological industries while boosting industrial output and income in the village had another significant impact on its economy: it was clear that, without careful management, bacterial residue from the DL *(ru suan)* and other bio-tech. factories would lead to water and soil pollution. To overcome this problem the decision was made, with help from experts in Shanghai, to develop an ecological system which would deal with the pollution and strengthen the overall economy of the village. Thus, it was primarily to overcome a potential pollution problem that the leadership decided to adopt an ecological agricultural system in 1993.

Traditional agriculture dominated in Qian Wei until 1992. As suggested earlier, individual household cultivation was replaced by *jiti* managed agriculture by the mid-1980s and the fields were soon worked almost exclusively by farmers hired from outside Qian Wei as the local villagers found employment in industrial enterprises, although decisions regarding what to produce, and how, were made by the village leadership. During the 1980s, change was gradual: the main cereals remained rice and maize, although outputs per *mu* grew as better strains, chemical fertilisers and multiple cropping - traditional 'Green Revolution' processes - were adopted. The village began to plant vegetables and fruits only at the start of the 1990s. In that the fields were not divided, the farmers were able to use machines for planting and harvesting. 1.1m yuan was invested in farm machinery, including eight large tractors, four other machines and mechanised irrigation. The village did not practise, had never heard of, ecological agriculture, however, although household level biogas pits had been experimented with, largely unsuccessfully, for seven or eight years in the mid-to-late 1970s.

The transformation of the industrial structure in 1992 became the catalyst for the transformation of the entire economy of Qian Wei (still, at that time, known as Sheng Li). Experts from Shanghai and elsewhere were invited into the village not only to advise on biotechnological processes but also on the establishment of an ecological agricultural system. The credit for this, as for much of the village's advance, is given to *shuji* Xu Weiguo who decided that if industrialization was to deepen, it was very important to nurture the natural environment. However it was in the same year, 1992, that a change of personnel in the leadership of the village occurred. Yang Pinjuan, a woman of 40 who had been 'doing CP

183

work' for the previous 5 years and who had been involved in the construction of the village since 1984 became *cunzhang* and with the village leadership came the responsibility for agricultural output. A year later, in 1993, Lu Zhenda was appointed from outside the village to be head of ecological agriculture and Shen Huichun became the new Party secretary *(shuji)* at the same time that Xu Weiguo became President *(zong)* of the newly constituted Shanghai Huaying High Tech Group Co., Ltd. Thus by mid 1993, Qian Wei *cun* not only had a new name and new administrative responsibilities for 2 neighbouring villages, it had a new leadership team, a new corporate structure and was committed to the development of CEA.

The establishment of a 1000 head pig farm in 1993 was one of their first acts, the pig farm playing a key role in the system. On the one hand, it was designed to absorb the bacterial residue from the DL factory, with stalks from the fields and leaves from the vegetable gardens to provide fodder for the pigs. On the other hand, the night soil of the pigs became the primary input for the new biogas digester built in the same year. Meanwhile, the pigs themselves began to provide an income for the village, mostly sold 'on the trotter' in the markets in Shanghai.

The biogas digester, the hub of the archetypal CEA system, was built in 1993, designed in Beijing on the basis of German experience, and began producing gas in May 1994. It was built at a cost of 1m yuan and by 1995 employed three full-time workers, with Lu Shanxi as its manager. It is one of the largest in the country: it has a capacity of 600 m^3 and is presently fed by two night soil pits (with a further two in reserve) each with a capacity of two tons of pig excrement! Whenever the pigs (and human residents) cannot produce sufficient input, cow dung is bought from outside the village to fill the gap. The biogas digester produces gas for cooking in every household in Qian Wei, except in the two small villages recently merged into the new *cun,* where stalks, grass and wood are still used for cooking purposes. In addition to energy for cooking, the biogas digester also powers a small 50KW electricity generator. Meanwhile, the slurry from the biogas digester is separated into liquid and solid waste, the former being used, via underground pipes, to fertilise the vegetable gardens and orchards, the latter used as feed in the fishponds.

A novel aspect of the system also designed to use by-products of the DL factory is the raising of baby turtles. Started in 1993, 1.8m yuan has been sunk into the project. Surplus steam from the DL factory is captured and is fed into the turtles' breeding houses to provide the necessary warmth. The turtles grow to an average weight of 5 kilograms, fed on fish and pigs' livers, and after 4 months are sold in the markets in Shanghai. In 1995, the anticipated output was 10,000 baby turtles and 8,600 adult turtles. Fish, fed on pig manure and biogas slurry, are also farmed: there are 550 *mu* of fishponds and 100 *mu* of turtle ponds, eleven ponds in all.

Agricultural practices have undergone considerable changes in the last two years as CEA has developed, in particular, fields previously used to raise cereals have been planted with fruits, including pears, apples, oranges, grapes and melons, vegetables grown inside very large polythene greenhouses, and flowers. Qian Wei now has 1280 mu of land devoted to vegetables and 450 mu to fruit, on which grow 10,000 fruit trees while the flower nursery is 1500m^2. In 1994, Qian Wei produced 2,000 tons of 'green' vegetables. The vegetables and fruit are fertilised predominantly from biogas slurry and are grown as 'green' fruit and vegetables.

There has been a conscious reduction in the use of chemical fertiliser and an increase in the application of organic fertilisers: at present the ratio of organic to chemical fertiliser used in Qian Wei is 70:30. Rice yields on the 600 *mu* of fields still used for cereals have risen to 1200 *jin* per *mu*.

As a result of the introduction of animal husbandry, the large biogas digester, turtles and fish and the growing of 'green' vegetables and fruits, Qian Wei has now completed the construction of an archetypal, modern, ecological agricultural system, albeit on the back of developments in biotechnology.

Related to Qian Wei's 'clean' industry and ecologically sound agricultural policies has been an ambitious decision to develop 'eco-tourism'. Trees have been planted to provide a forest belt 400m long and 33 m wide alongside the fields, but a further 4,000 trees of different varieties have been planted in 45 *mu* of specially created 'scenic' areas and parks. Altogether, forests now cover 15 per cent of the land area. Newly and elegantly painted pagodas have been built amongst the ponds, traditional style bridges have been constructed over them and a small museum has been established. Meanwhile, a traditional mediaeval village has been built in a field close to the modern village centre. Karaoke and dancing rooms have been established, as has a restaurant and two comfortable guest houses for visitors. According to Zhao Weidao, the village has already received tourists from Shanghai, although the majority of the visitors have come essentially 'to inspect and learn from the economic developments in Qian Wei.' This is hardly surprising: though many of the new buildings are pleasingly cheerful and much of the land is thoughtfully landscaped, Qian Wei is, as yet, no great tourist attraction. Allied to the development of tourism in Qian Wei itself, the village owns and manages a hotel in Shanghai, the San Shan Hotel, providing additional income to the *jiti*, and the building of a further hotel on the northern border of Qian Wei, by the sea, is planned.

8(6) 5. Political Economy in the 1990s

8(6) 5.1. Political structures

The leadership of Qian Wei, provided wholly by the Chinese Communist Party branch, comprises seven people, including Yang Pinjuan, the female leader of the village appointed in 1993 at the age of 41, who has, in addition to other duties, particular responsibility for agriculture, Shen Huichun, *shuji* since 1993, two deputy *shujis* one responsible for industry and another for relations with Shanghai, the manager of the FE toothpaste factory and the leader of the Victory shoe polish factory. *Zong* Xu Weiguo still commands complete authority, however, and all the main enterprises, including agriculture and animal husbandry are coordinated within the Hua Ying Group Company of which he is president. The leadership meets formally once a month and makes all strategic plans; additionally, villagers and each of the main factories have representative committees. In 1994, there were 74 members of the Communist Party in Qian Wei.

Although renowned as an ecological agricultural village today, Qian Wei earns the majority of its income from its high-tech industries. The (third generation) FE toothpaste is the most important of these, with 150 workers producing an annual output of 80m.yuan and yielding an annual profit of 15m. yuan in 1994. Fortunes have fluctuated fast, however, to the extent that profits fell to only 0.8m yuan in 1996, blamed on the difficulties for such a small company to compete nationwide in marketing terms against the foreign multinationals entering the Chinese toothpaste market. (*Shuji* Shen Huichun was pleased to hear that I had seen FE toothpaste in Beijing shops, however). The Hua Ying Group is now planning cooperation with a Malaysian company with the intention of further product development and entry into international markets.

The next most important industry is the manufacture of Victory shoe polish, while other products include car polish, electronic components (light switches and lamp standards), glass products, adhesive tape and packaging, the latter factories recently set up on the northern boundary of the village. The contribution of DL *(ru suan)* , despite employing 130 workers mostly from Qian Wei, however, has still not come up to expectations. Qian Wei's highly automated factory is the only one making the product in China, (a product which was mentioned in China's Eighth Five Year Plan as being an important product for the future) but there are still problems in its development and application and its potential is not as yet reflected in profit, its domestic market being presently too small. However the leadership still believes it has a bright future and the Hua Ying Group has recently signed an agreement with a Dutch company making the same product (the president of which visited Qian Wei earlier in 1997) and jointly they have ambitious plans to export both to Japan and to European countries.

The most profitable recent development has been that of aquaculture, with expansion of the raising of 'baby crabs', turtles and eels. 350 kilos. of 'baby crabs' are artificially hatched every year (there are approx 150,000 baby crabs per kilo.), while the village raises 100,000 river eels and in 1997 70,000 turtles. According to *shuji* Shen Huichun, aquaculture promises 'new directions' for the village, particularly so because there has been a dramatic reduction in the natural hatching of crabs and raising of eels at the mouth of the Yangtse River. In 1996, the profits after tax from aquaculture were 2.5m yuan, the most profitable enterprise of that year.

Over 1,000 workers are employed in all in Qian Wei, half of whom come from outside Qian Wei and many of whom live in dormitories. All Qian Wei villagers work in the factories, the fields being worked (apart from the manager) by farmers from Zhejiang province who live in older houses on the northern border of the village. Qian Wei villagers, apart from outside experts and specialists, are paid more than those employed from outside. Pay varies a little with the degree of responsibility, but the average salary for villagers is 6000 yuan a year, 50 per cent more than outsiders can expect. Managers earn 'only a little more' than workers, their income depending upon bonuses tied to profits. Income per head in 1995 was approx. 4000 yuan, comparing favourably with an income per head of 3000 yuan across Chongming Dao as a whole. In 1996, income per head was 'around' 4,500 yuan, making Qian Wei the second richest eco-village (behind Teng Tou) researched, on a par with Liu Min Ying.

From very recent, very humble beginnings, Qian Wei is now one of the foremost villages in the Shanghai area, with a strong industrial base, animal husbandry and aquaculture, a developing tourist industry and a very modern ecological agricultural system. Every year from 1984 to 1993, it won a 'civilized village' award (in 1988, receiving the award from Jiang Zemin) and in 1994 was named 'model village' by the Shanghai government. It has over the years received many visitors, including current Premier Zhu Rongji when he was mayor of Shanghai, and Song Jian, currently State Councillor with responsibility for environmental protection.

8(6) 5.3 Social developments

As the economy of the village has developed in the 1980s, so has its social infrastructure. When the village was first created, houses were very basic and villagers slept and cooked on the traditional *kang*. Now, most of the villagers with their *hukou* in Qian Wei live in three-storey houses, the rest in two-storey houses, where average floorspace per head is 70 square metres. The houses, all brick-built within the last ten years, are very pleasantly designed, with unusually sloping roofs, in a Dutch style, and are big by any Chinese standards. All have electricity and water (supplied from deep underground wells), biogas for cooking provided at a token charge, and most have recently installed telephones and televisions hooked up, amongst other channels, to Star TV from Hong Kong. The residential area is in the eastern part of the village nucleus, bounded by walls and linked to the other parts of Qian Wei by well maintained white concreted roads. Altogether, there are 8,200 m. of white concrete roads and 1,500 m.. of walls in Qian Wei. In 1992, at a cost of 1.95m yuan, a road was built from the south gate of the island (the ferry port) to the village gates.

Qian Wei has its clinic and own kindergarten and there are plans to open a primary school in 1997. According to Yang Pinjuan, there are now no difficulties in implementing the one-child birth control policy since most couples don't want more than one child, given the increasing living standards associated with factory work and the economic sacrifices that a second child implies. Uniquely, Qian Wei established a cemetery in 1995.

8(6) 5.4 Environmental developments

Qian Wei remains a very ambitious village ecologically. In propaganda material produced by the village (undated, unauthored), it lays out its desire to create eight 'ecological systems' *(xitong)*: (i) an 'ecologically clean vegetable system', on 1280 *mu* of fields, producing an output of wholly 'green' vegetables of 2000 tons a year, (ii) an 'ecologically specialised raising system', with 550 *mu* of fishponds, and 100 *mu* of turtle ponds. In 1995 there were already 1600 m^2 of ponds for baby turtles. Planned production is an output of 70,000 baby turtles and 40,000 adult turtles annually, (iii) an 'ecological grain system', with rice output of 600 kilos per *mu,* on 600 mu of fields fertilised entirely organically from stalks and biogas slurry, (iv) an 'ecological fruits system', with 450 *mu* planted with fruit trees, making gardens for every household and

187

protecting the environment for birds, (v) an 'ecological factory system', with the DL lactic acid factory and the FE toothpaste factory at its core, (vi) an 'eco-residential economic system', where domestic waste products and human night soil are fed into the communal-level biogas pit, (vii) an 'eco-tourist system' with scenic spots, hotels, fishing, karaoke,and (viii) a 'spiritual civilisation system' with dancing halls, a kindergarten, a primary school, a library, and television and telephone in every home.

By observation in 1997, Qian Wei has already progressed a long way in this direction. Agriculture involves 'the three highs: high input, high quality, high profit' (meaning, by high input, application of (mostly) organic fertiliser, use of machines and advanced irrigation). A farm producing 'green' vegetables and fruits supplying Shanghai with 600 tons of green food has been established (the farm in 1997 was in the process of getting its products approved by the Green Food Agricultural Bureau in Shanghai in order that its products can be labelled with the Green Food sign). Forestry, with 370,000 trees planted by 1997, has been developed. Animal husbandry and aquaculture have expanded impressively, the biogas digester, despite problems of distribution resulting from poor laying of pipes to some households, still provides energy for cooking to most households and is the centre of an archetypal CEA system, supplying the slurry to fertilise the green vegetables. Meanwhile, impressive strides have been made in developing eco-tourism: the mediaeval village in particular has been very tastefully and imaginatively designed.

8(6) 6. Conclusions

Qian Wei has transformed itself twice in the last 30 years: reclaimed from the sea in the late 60s and early 70s, it steadily developed its agricultural and industrial strength for 20 years in fairly conventional ways before it was forced by the collapse of demand for its toothpaste in the early 1990s to rethink its entire economy. The outcome is now a very modern village with an industrial structure biased towards science-based bio-technological processes and a non-industrial sector dominated by animal husbandry, aquaculture, the production of 'green' vegetables and fruits, and eco-tourism. The ecological agricultural system consciously constructed with a community level biogas digester at its core, is a textbook model. Waste from the factories and stalks from the fields are used as fodder for the pigs whose night soil is used as input to the biogas digester. The latter provides clean power for cooking, while its slurry is used either in liquid form as fertiliser for the vegetable plots and rice fields or in solid form as fish food. The fish are variously sold or fed to baby turtles, themselves reared from surplus heat recovered from the factories. Meanwhile, the soil quality has steadily strengthened with the progressively increased application of organic, biogas slurry rather than chemical fertilisers. This system made Qian Wei one of the favourites for the UNEP 'Global 500' award in 1996. The village has been awarded plaudits from the county and from Shanghai for both its industrial and ecological agricultural advances.

Qian Wei has made sufficient economic and environmental progress to be used as a model of development by the local Chongming Dao Environmental Protection Bureau, set up in 1994 and challenged to develop ecological agriculture by the designation of Chongming Dao, also in 1994, as one of NEPA's 50 'ecological counties'. Huang Yongzhou, project officer in the bureau, in common with Zhao Weidao, industrial consultant to Qian Wei, argues that the primary factor determining the successes of the village to date has been the quality of the leadership, particularly that of the young, early leader and *shuji*, Xu Weiguo, now President of the Hua Ying Group and the current leader, Yang Pinjuan, both of whom in earlier years 'devoted enormous energies to village construction' and who in recent years 'have paid great attention to environmental ideas and the principles of ecology.' According to Zhao Weidao, 'the leaders were never satisfied and never took anything for themselves.' Huang Yongzhou puts special emphasis on the leadership given the fact that Qian Wei had not originally been a rich village blessed with resources, rather the reverse. Yang Pinjuan herself suggests that the primary secret of Qian Wei's success has been 'leaders working hard together with the same purpose as the villagers.' In 1997, when I was shown around the village and enterprises by the now 45 year old *zong* Xu Weiguo himself, it was tangible that he was held in enormous respect and affection by ordinary villagers.

Another factor which has clearly militated in favour of Qian Wei has been its organisation along *jiti* lines. Yang Pinjuan suggests this has been 'very important', particularly in the development of ecological agriculture. The present modern ecological agricultural system, including the biogas digester, was very expensive to set up and, according to Zhou Weidao, the initial expenditure was not immediately popular with the villagers and could not have been achieved in a village broken up into individual households. This is true partly because such villages would find it difficult or impossible to raise sufficient collective funds for the investment, partly because it would be more difficult to develop a collective will to execute it. Today, it is agreed that the system is now popular with the villagers who have been attending training classes from experts from Shanghai and Beijing.

That the village is run on *jiti* lines is largely the result of the early industrialization and the imagination of *shuji* Xu Weiguo in the early 1980s who was keen to promote the rapidest possible economic advance and thus promoted, as farmers voted with their feet as they moved from field to factory, the rapid amalgamation of the fields only recently decollectivised. Today the *jiti* operates as the Hua Ying Group Company but the change is merely to form and not substance. Each enterprise operates as a separate entity within the Group with a relationship analogous to that of a 'parent and child.' Xu Weiguo himself is adamant that the *jiti* has been crucial in the development of agriculture generally and CEA in particular. He says the immediate recollectivization meant that the farm machines and basic infrastructure were immediately available. Mechanization of farm operations has continued on a firm base. The *jiti* was also important, he says, because many individual farmers simply weren't very competent in growing crops.

Despite being cut off by water from the great metropolis of 15m. people, Qian Wei has taken considerable advantage of its proximity to Shanghai in two primary respects. Firstly, being close to a large centre of population has meant the village has been able to integrate itself into important local markets, both for the supply of its raw materials and as a destination for its products. For example, it sells its pigs, baby turtles and 'green' vegetables and fruits in

Shanghai and it is unlikely that it would be feasible to consider eco-tourism without such a location. Secondly, Qian Wei has never been afraid of making best use of finance and expertise to be found there. The original toothpaste factory was set up in conjunction with a Shanghai enterprise bringing money and management skills while the new, thriving bio-technological FE toothpaste factory, and other high-tech enterprises, have been set up in conjunction with experts from Fudan University and other scientific institutes. It is clear, unlike the boast of many eco-villages, that Qian Wei has *not* done it on its own, that outside experts and consultants have been and continue to be employed in Qian Wei. Many such experts, including Zhou Weidao, who came in 1989 as an industrial consultant, and has become so enthusiastic for the development of Qian Wei, that he has never left, live permanently in the village. Presently, there are five such experts. And Qian Wei's recent opportunities of attracting foreign investment clearly derive partly from its favourable location.

Qian Wei has also benefited from good *political* relations with Shanghai. Throughout the 1980s it won awards from Shanghai government and received high-level visitors, such as Shanghai's then mayor, Zhu Rongji. Shanghai has also provided Qian Wei with financial assistance: the biogas digester, for example, cost 10m. yuan, half of which came from the Shanghai City government. The road from the south gate of Chongming Dao to Qian Wei was also built, at a cost of 1.95m. yuan, by Shanghai. As with many enterprises, success has seemed to breed success: having made a name for itself through its own devices in the late 70s and early 80s, it increasingly came to the attention of the county and Shanghai authorities and as a result gained the benefits of its patronage.

While Qian Wei's ecological agricultural system is modern and comprehensive and is an exemplar of successful CEA generally and while that system was the basis of its eventually unsuccessful claims to the UNEP 'Global 500' award in 1996 (no awards were made by the UNEP to Chinese eco-villages in that year), the village's economy is overwhelmingly skewed in favour of industry which contributed 88 per cent of the village output of 100m. yuan in 1995. Indeed, it is clear that in the case of Qian Wei, ecological agricultural development followed industrial development, indeed was effectively *dependent on* and *determined by* it. It is extremely unlikely that CEA would have been adopted had it not been for the need to regenerate industrial activity in a non-polluting manner. The fact that very recently, Qian Wei has become industrially very prosperous allowed the leadership to introduce CEA despite initial lack of understanding of it amongst villagers. In 1997, Qian Wei's motto is 'develop the economy, protect the environment.' (promotional video, 1997)

In summary, ecological agricultural development in Qian Wei has benefited from strong, imaginative and trusted leadership, outside assistance resulting from its relations with Fudan University and other scientific institutes, good relations with political leaders in the county and in Shanghai, good transport and telephone links with markets, a prosperous industrial economy and *jiti* organisation. Each of these features has played a part in ensuring its current advances. However its development has been essentially entrepreneurial and its high level of material advance so far achieved has been predicated on the successes of relatively high-tech and risky industries. Though Xu Weiguo extols the communal virtues of Mao Zedong and continues to live in an ordinary house in Qian Wei, he operates less like a socialist political leader and more like a hard-headed businessman, as he journeys backwards and forwards from Qian Wei to Shanghai in his chauffeur-driven black limousine (I got a lift to Shanghai at the end of my 1997 visit) doing business deals, continuously engaged on his mobile phone. While he remains successful in those business deals, then Qian Wei, under the guise of the Hua Ying Group Company, will continue to prosper and CEA will do likewise. But it is clear that Qian Wei is taking a road which is not risk-free.

8(7) Tie Xi

8(7) 1. Introduction

Tie Xi is a small village in the suburbs of the town of Mishan, two kilometres from its centre and in the county of the same name, in the south-east of China's most north-easterly province, Heilongjiang, and but a few kilometres from its eastern border with Russia. Mishan itself is very inaccessible: to get there (if you can get hold of a ticket) involves a twelve hour train journey of maximum discomfort from Harbin, Heilongjiang's provincial capital, which is itself an eighteen hour train journey from Beijing. Heilongjiang is an overwhelmingly agricultural province, with 80 per cent of its population engaged in agricultural occupations, including animal husbandry and forestry, with 70 per cent forest coverage, but it is also mineral-rich and has substantial coal, oil and mechanical engineering industries. Despite plentiful resources, however, its cadres argue that Heilongjiang is not a wealthy province because of the much higher levels of state intervention there than is common elsewhere in present-day China's market orientated economy: there is considerable state ownership and regulation of industry and product prices are strictly controlled. It is claimed by the province officials that because of the greater availability of resources in Heilongjiang than in other provinces, the latters' development depends on resources being transferred from Heilongjiang. Crops produced in Heilongjiang, for example, are bought by the state at low prices, keeping farmers badly off, and transported to other, often richer provinces.

Tie Xi has a population of 1640 people living in 420 households and has a distinctive identity and territory within Mishan, being to the west of the railway line stretching into the open countryside, although it is on the west edge of Mishan, and merges into it to all other intents and purposes. Mishan County has a total population of 420,000 engaged mostly in agriculture and in textiles, coal mining, edible oil manufacture and food processing. Because of its close proximity to the Russian border, there are many Russians in Mishan engaged in border trade: Russians will buy food, clothing and materials from China and sell chemicals, chemical fertiliser and industrial materials. (Almost all foreigners are Russian: it was initially assumed by everyone that I was one).The Environmental Protection Bureau in Mishan opened in 1979 but the still high and evident levels of atmospheric pollution caused largely through coal burning suggests it has an uphill task. On the short journey from the EPB office to Tie Xi, and just next door to the latter, lies a cement factory which furiously belches out black smoke from its chimneys day and night, hardly appropriate to Tie Xi's reputation as an environmentally friendly place.

While the nucleus of Tie Xi's residential and industrial area is compact and clustered around the main village office buildings, its landholdings are extensive, owning as it does huge tracts of arable land, forests and enterprises far away from the village itself. It is more like the headquarters of a huge, geographically dispersed business empire than a village. From very unpromising origins it has become Heilongjiang's 'first village' and in 1995 was being pushed for nomination as a candidate for the UNEP's Global 500 award by Heilongjiang Environmental Protection Bureau for 1996.

8(7) 2. Recent History

Tie Xi is of very recent origin, founded in the dying days of the Cultural Revolution in early 1976 when on March 16 there was a reorganisation of Heping brigade *(dadui)* and as a result three production teams *(xiaodui)* combined to form Tie Xi brigade. The name was chosen because of the location of the new brigade: west of the railway line. At its inception, there were 292 households and a population of 1108 in a total area of 2,614 *mu*. At the time, it was in a parlous state financially, with a reputation for poverty. It was said to be a brigade of 'two haves and three have nots *'(liang you, san wu)* : it had collective debt of 240,000 yuan and individual debt of 210,000 yuan, yet it had no capital, no enterprises and no security of living (propaganda sheet, undated, anon, given to me by Mishan EPB, 1995, in Chinese). Average income per capita was 80 yuan.

A new Party branch was set up and Zhang Hongsheng, with 2 years education in vocational college, was elected Party secretary *(shuji)* with a determination to turn the situation around. After many meetings and discussions, the leaders came to the realisation that they couldn't get out of poverty merely by extracting food from the soil, that it was necessary to 'seek fortune beyond the soil.' They decided that the way to prosperity was to develop sideline enterprises. Given that the Cultural Revolution had not yet come to an end, there was plenty of criticism levelled at the leadership for 'taking the capitalist road'. But the leadership stuck to its guns. They set up a production team, began quarrying in the hills and set up a brick-making factory - the first enterprise in Tie Xi - in late 1976. The first batch of production from the factory was a failure, however, and the 10,000 yuan investment in it, financed from a bank loan, went down the drain. Undeterred, a construction team was established and a building project in Mishan undertaken. The project earned sufficient money to pay off the 10,000 yuan loan and to buy four horse drawn carts, the latter allowing the first transportation team to be established.

In the spring of 1977, 65 men formed the sideline production brigade which engaged in quarrying, construction and transport. Within 2 years the production brigade earned 650,000 yuan, allowing Tie Xi to enjoy a per capita income of 200 yuan. The brigade leaders continued to be criticized as 'capitalist roaders', even in 1978, but *shuji* Zhang Hongsheng was undeterred, arguing that 'there was no hope for socialism if the people remained poor.' The village took work from the Provincial Forestry Bureau working a forest farm some distance from Tie Xi. The work team endured very harsh conditions, ate coarse food (and drank cold water) but made 230,000 yuan which was reinvested in the brick-making factory back in Tie Xi. Experts were invited to Tie Xi to oversee operations. In 1979, the factory manufactured 900,000 red bricks with a production value of 60,000 yuan: Ti Xie's first village-run industrial enterprise was by then a success and the village was turning the corner as far as its economic difficulties were concerned.

In April 1979, 200,000 yuan was invested in establishing animal husbandry in Tie Xi: a pig farm and a chicken farm were established and together made a profit of 90,000 yuan in their first year of operation. In September 1980, the village bought some land in Mishan County with 15,000 yuan and established a service building with a hotel, restaurant and shop. In September 1982, a fish farm was set up, in 1983, a freezer storage unit was established and in 1984, in partnership with Bei Xing agricultural farm, an open-caste coal mine was opened, approximately 100 kilometres from Tie Xi itself. Later in 1984, a chemical fertiliser factory was established. All of these enterprises were not merely 'firsts' for Tie Xi, but firsts for Mishan County as well. It was commonly said at the time (reportedly) that the leaders of Tie Xi were 'ahead of their time.'

However, while this rush of industrial, commercial and animal husbandry enterprises changed the immediate economic circumstances of Tie Xi for the better, the leadership also embarked upon a venture with arguably more long term significance. During the Cultural Revolution (1966-76) a forest farm had been set up in Mishan County, 45 kilometres from Tie Xi, as a training ground for youth sent from the cities for reeducation amongst the peasants. At the end of the Cultural Revolution, it was still being worked by 'city youth', although many had deserted, the forest management was disorganised and the trees, where they had not been cut down, were in a poor condition. By 1981, Mishan County was keen to offload this problem, particularly so given the concern expressed at the time by officials over the extent of the deforestation that was taking place as forested land, subsequent to the rural reforms which were taking place at the time, was being divided up and 'privatized'. As a result Tie Xi village 'bought' the forest farm from the County and undertook to work and manage it. So, in April 1991, a forest farm with adjacent arable land of 16,500 *mu* in total was added to Tie Xi's armoury of enterprises.

(7) 3. Land Reform

The rural reforms, ushered in by the decisions made by the 3rd Plenum of the 11th CP Central Committee in December 1978, posed no particular conflicts for the Tie Xi leadership, committed as they were to a more marketized and enterprise economy than was the prevailing wisdom at the time, particularly in Heilongjiang, a traditionally conservative province, and in Mishan itself, from which the leaders had earlier received criticism for taking 'the capitalist road'. Indeed, it was suggested to me by villagers that in the late 1970s the county was in the hands of leaders who favoured 'ultra-left policies' and that the village leadership was very strong and resourceful in establishing enterprises in the teeth of them.

In 1982, a considerable amount of discussion took place amongst the leadership and in village-wide meetings and a consensus was reached on the basis of *shuji* Zhang Hongsheng's advice with regard to the division of previously collective land and property: according to Zhang Xingbo a 62 year old ex-school teacher, 'those lands considered right for division were divided, those considered not right for division were not.' This meant that the fields close to the village were divided amongst households (effectively becoming *kouliangtian*), the land far away from the village, including the recently acquired forest farm and adjacent arable land was not. Additionally, all the property of the various industrial and enterprise teams, which at the time included the brick factory, construction and transportation teams, the chemical fertiliser factory, the cattle and chicken farms and the service building, were kept in the hands of the collective, a common enough outcome in rural China.

As a result, each household was allotted, on the basis of 1 *mu* per head, land which averaged 3.5 *mu*. The plots were officially leased to the villagers for 15 years while ownership remained with the collective. According to the villagers, the plots were planted almost exclusively with vegetables which were supplied to Mishan County, the main staple crops (wheat, corn and rice) being provided through procurement by the state. The land division was popular, largely because the villagers felt they could make more money as a result of it. Attitudes have not changed: the land division amongst households is taken-for-granted; meanwhile growing vegetables and selling them to Mishan have become largely part-time occupations allowing villages to engage in a range of small scale entrepreneurial business activities *(geti hu)* or gain employment in a collective enterprise.

8(7) 4. Chinese Ecological Agriculture

Ecological agriculture has been adopted in Tie Xi in an extremely gradual and pragmatic manner and a pure ecological agricultural system per se has never been adopted. What ecological agriculture is practised takes place most consciously on the arable land close to the forest farm, in a semi-mountainous forested area some 44 kilometres north of Tie Xi village (bought from the county in 1981). It is suggested by the propaganda sheet from the Mishan County EPB that the village 'began along the road to ecological agriculture with a belief in the importance of developing economic resources by creating a "virtuous circle" through a process of recycling within a broadly comprehensive system of rural production, with forestry at its core.' In this way, it was argued, environmental, economic and social benefits could be integrated.

From the early 1980s on, the expanded planting of cereal crops (rice and wheat) took place alongside the establishment and integration of animal husbandry, fish ponds and forestry. The planting of trees to build up the forest belt was the first priority, then came the raising of animals and the digging of fishponds. A base for growing edible fungus *(mu er)* was established. In the early 1990s the Tie Xi Forest Farm Eco-complex Natural Conservation Area was named and established, covering 72.35 km2. The farm in 1993 consisted of 810 hectares of forest, of which 260 hectares were man-made, a river (the *Ta Tou* river), a small reservoir with a carrying capacity of 261,000 m3., 650 *mu* of fish ponds on which ducks and geese were raised, 1,500 *mu* of arable land and 150 *mu* of garden nurseries. The water is non-polluted (it certainly *looked* clean in 1995) and reportedly has a high Ph value, good for fish growth. According to the village propaganda the development took place 'under the guidance of the principles of ecology and environmental economics' to encourage 'low input, high output production.' As more animal nightsoil became available, the application of chemical fertiliser and pesticides to the fields was reduced to a minimum, improving soil quality and (reportedly) increasing cereal yields. At the same time, the forest cover reduced soil erosion, slowed water loss and improved the micro-climate.

The villagers in Tie Xi itself practice ecological agriculture only to the extent that households have been encouraged to raise animals to provide night soil as fertiliser for their largely vegetable plots. When I asked villagers (Lu Xunwen, Guo Shurong and Zhang Xingbo) about ecological agriculture, there was no obvious sense of recognition and there was certainly no conscious marketing of 'green' or 'organic' vegetables. Chemical fertilisers were used, at least in part, and so were pesticides. There was no biogas production of any kind in Tie Xi in 1995 and no plans to develop it.

It was claimed, somewhat ritualistically by Zhang Hongtian, that China's national one-child policy was strictly adhered to. The average number of persons per household (roughly 4) suggests otherwise, however, as did the testimony of the good-humoured driver of the Jeep, who told me he had two children aged 6 and 2, and that such a state of affairs was very common. He also told me he had no interest in environmental protection.

It is the all-round development of the Tie Xi Forest Farm which has brought Tie Xi to the attention of Heilongjiang's environmental protection officials and forestry is its major claim to fame. Rare flora and fauna abound: while the water is important in encouraging certain wildlife including fishes, snakes, amphibians, geese and ducks, including seven pairs of mandarin ducks the forest cover is seen to be important to the preservation of wild animals. In 1993, according to the propaganda, wildlife there included 2 black bears, 21 red horse-deer *(ma lu)* and large numbers of ordinary deer *(mei hua lu)* . In 1993, the total value of output of the farm was well over 1m yuan: income from animal husbandry was 225,000 yuan, from fisheries 225,000 yuan, from agriculture, 400,000 yuan and from forestry itself 500,000 yuan. The latter figure is not high by relative standards yet, but, it is claimed by the leadership, is a store of wealth for the future, a 'green bank.' Zhang Hongtian was very firm in his view that trees were 'good for the environment and good for the economy.'

8(7) 5. Political Economy in the 1980s and 1990s

8(7) 5.1 Political structure

The village had 'collective leadership', with the village committee composed of the *shuji,* Zhang Hongsheng, deputy *shuji* and *cunzhang,* Zhang Hongtian, brother of Zhang Hongsheng, and the leaders responsible for the main industrial units, birth control, public security & land management. Zhang Hongtian took it for granted that all would necessarily be members of the Party: indeed he made it clear that all important decisions had to pass through the Party branch committee for ratification (my questioning concerning the overlap of the village leadership and Party committee made him exasperated at my lack of awareness of 'how it is done' in China). Clearly, politics are still strong in Tie Xi: further propaganda material (dated 1993) boasts of the numbers of villagers who are Party members (107) and Communist League members (289) and the room in the main office block given over to relating the recent history of the village through the written word and photographs explains in bold letters that Tie Xi is a 'prosperous, socialist village under Communist Party leadership.' It does not mention ecological agriculture.

8(7) 5.2 Economic developments

While Tie Xi is formally a village *(cun)* within the county of Mishan, and while leaders talk in term of collective *(jiti)* organisation, for purposes appropriate to China in the 1990's, it has renamed itself the Tie Xi Enterprise Group Corporation *(Tie Xi Qiye Jituan Gongsi)* with Zhang Hongsheng as Managing Director and Zhang Hongtian his deputy. This organisation effectively owns and manages all the collective enterprises of Tie Xi, including its industries, animal husbandry, forestry, fishery and collective agriculture. Only household agriculture and small-scale entrepreneurial activities are excluded. It has 8 'intensive' and 3 'extensive' branch companies, with 48 enterprises altogether and, with fixed assets of over 50m yuan, employs 540 employees, many from outside Tie Xi. Its main businesses includes chemical fertiliser production, construction and installation, the coal mine, transportation, forestry, commerce, boundary trade, agriculture, animal husbandry and fisheries and food processing. As is the case in many other parts of rural China, the village has turned itself into a holding company for all the non-household businesses enterprises in Tie Xi, with the village leadership as its board of directors. In 1994, the total value of the Corporation was 101.36m. yuan, surpassing 100m. yuan for the first time, up from 74.58m yuan in 1993. (These are nominal, not real values. Inflation in China between 1993 and 1994 was about 20 per cent.) The Corporation provided over 80 per cent of the income of the village. According to Zhang Hongtian, annual income per head in Tie Xie was a little over 2750 yuan in 1994.

The major industrial income earner within the Corporation is the open-caste coal mine 'allocated' to Tie Xi by the Mishan Mineral Resources Agency, opened in 1984, and which, according to Yan Tiezheng (chief of Mishan EPB) in 1995 employs 250 miners and produces 230,000 tons of coal with an output value of approximately 2.5m yuan. The mine, 100km from Tie Xi, is owned and managed by the village, although it is worked by miners from villages close by. Other impor tant enterprises located in the village itself are the chicken farm with 40,000 laying chickens, the pig farm with 10,000 pigs and the chemical fertiliser factory, all of which were undergoing expansion in 1995, indeed a brand new chemical fertiliser factory was being built in 1995, with a capacity of 60,000 tons, six times the capacity of the present one. Another new venture for the Corporation is a plant producing mineral water, built in 1995, planned to open at the end of that year, on open, rural land 20kms from Tie Xi (on the way to the forest farm) jointly owned by Tie Xi and Mishan power supply bureau. The mineral water emanates from an underground spring beneath the factory.

Tie Xi's economic development has allowed the living standards of the villagers to rise considerably over the years. In 1976, annual income per head, at 80 yuan, was poor even by the standards of the locality, by 1984, it stood at 2,750 yuan, higher, according to Yan Tiezheng of Mishan EPB, than any other village in Mishan and one of the highest in Heilongjiang. Villagers have reached a 'comfortably well-off' standard of living *(xiaokang)* well ahead of many other Chinese and well before the designated year 2000.

8(7) 5.3 'Social' developments

In common with villagers across many parts of China, private household consumption has expanded dramatically, so that villagers enjoy not only a vastly more varied diet to include vegetables, meat and fish on a routine basis but also they are likely to own colour televisions, video recorders and a range of consumer durables unimaginable even ten years ago.

Tie Xi boasts an impressive record of house building: in 1995 all households were mostly two storey brick-built with indoor cooking and toilet facilities, a long way from the mud and straw houses extant even in the early 1980s. The village has recently completed a three storey primary school building with 1,400 m^2 of floor space and a home for old people with 600m^2. The main village building, with 1,200 m^2, houses offices, a restaurant and an entertainment centre for the villagers and outside is an open dancing area. In 1993, there were 180 telephones in the village, 43 motorized commercial and agricultural vehicles and 12 cars. The village is proud to own cars made in Japan. I was taken around in a four-wheel drive chauffeur-driven Japanese Jeep.

8(7) 5.4 Environmental developments

While Tie Xi claims to have adopted ecological agriculture it is primarily its forestry which has brought it to the attention of environmental protection officials. In that so much deforestation took place in the 1980s in Heilongjiang, the development of forestry on the expanded forest farm has been a particularly impressive achievement, leading to improved soil quality, with reduced soil erosion and lower water loss, an improvement in the micro-climate, and protection for flora and fauna, and with regard to the latter, a protective environment for wild animals, including bears, tigers and deer. The forest is likely to provide opportunities in the future for 'eco-tourism', according to Piao Haidong; already a lodge has been built with good views over the forested area (suitably equipped with bedroom accommodation, a restaurant and the ubiquitous karaoke). Meanwhile the forest has begun to yield some income to the village, and with careful management, has the potential to become a 'green bank.'

Ecological agriculture per se, and particularly in its purer form, however, has taken rather a back seat. It seems little more than a bolt-on, an almost optional extra. In my interviews with village leaders and with villagers, it was clear that ecological agriculture was not an important concept:certainly more organic fertiliser was being used than before, but while there was some recycling of resources based on the development of animal husbandry and agriculture and some development of related sideline industries, particularly food-processing, it was clearly not the case that Tie Xi was a model *ecological agricultural* village. A village with rapid economic development, yes, and one with an impressive concern on both economic and environmental grounds to develop its forestry. And one that was aware and interested in the natural environment - Zhang Hongtian explained that the village leadership prohibited the establishment of polluting industry. But it was not a village for which CEA was paramount. Indeed, it was quite clear from the frustration shown by Zhang Hongtian at my persistent questioning about ecological agriculture in 1995 that this was the case and villagers that I questioned were not even aware of the concept. It was left to the Heilongjiang environmental officials and the rather vague Mishan County EPB propaganda leaflet to come to my aid to explain how it was supposed to 'work' in Tie Xi. And Piao Haidong (chief of the Heilongjiang Provincial EPB) quite openly admitted in 1995 that while there were experimental ecological agricultural sites in Heilongjiang, some even with biogas as a centre, Tie Xi still had some way to go in developing its ecological agricultural system and that its record on developing its *forestry* alongside the development of the economy as a whole had brought it to the attention of his office and was the reason for its nomination to NEPA as a contender for the UNEP Global 500 award.

8(7) 6. Conclusions

Tie Xi has clearly achieved very fast rates of economic development since its inception as a brigade 20 years ago and operates an unusual economy in that many of its economic activities and fixed assets (e.g. the forest farm, coal mine and mineral water plant) are situated very far from the village itself. It is like the head office of a corporation with an empire of productive assets far and wide. It has a thriving collective economy as well as a thriving private *(geti hu)* one. It has a very substantial range of economic activities, one of which is the Ecologically Complex Natural Conservation Area. According to Piao Haidong, there is no village like it in China, it is 'unique'. Meanwhile, economic development has delivered rapidly rising standards of living, both in terms of the private consumption of households and the public facilities available.

Tie Xi is located in an area of outstanding natural beauty, but its far-flung location on the far north-eastern border of China has given it no advantages in its economic and environmental development: it is certainly in a highly inaccessible location for a model village and has, up until now, received no help from any external agencies nor benefited from proximity to a major centre of population. It began life as a very poor village only twenty years ago.

The dynamic, enterprising and imaginative leadership of the two key leaders and brothers, Zhang Hongsheng and Zhang Hongtian seems to be the primary explanation for the particular successes that Tie Xi has achieved. This explanation was held by everyone I spoke to, whether it was the village leaders, villagers of Tie Xi or the environmental protection officials of Mishan or Heilongjiang. It is clear that it has been their entrepreneurial talents which have marked them out for particular honours: Zhang Hongsheng, *shuji* since 1976, has won numerous county and provincial awards as a model worker and model peasant entrepreneur in the 1980s and 1990s and his brother, and two other leaders, Sun Maoyi and Zhang Xinyou, both managers of branches of the Tie Xi Enterprise Group Corporation are acclaimed both inside and outside the village as key forces in change. Zhang Hongtian claims that the development of forestry was his brainchild, that he had understood the value of the forest as a teenager in the 1960s but that it was only after 1976 that he was in a position to do anything about it. Meanwhile, the Mishan and Heilongjiang officials make it clear that outside assets were increasingly allocated to the village by the county and province - e.g. the forest farm, the coal mine, the mineral water spring and factory - because of the growing reputation of the village leaders as efficient entrepreneurs and managers.

In conclusion, Tie Xi became Heilongjiang's main candidate for the UNEP's Global 500 award in 1996 and its candidature was based on its record of developing forestry within an expanding and ever richer economy. Obeisance to ecological agriculture has, arguably, been a necessary adjunct to its claims to be advancing on an environmentally friendly front, but it has not been a major guiding concept in the village's development; meanwhile, it has recently opened a *new* chemical fertiliser factory and the inside of the chicken farm suggests there is little sentimentality towards fellow animals. Following the theories of CEA propounded by NEPA and other authorities, the strong role for the collective *(jiti)* in managing the forest farm, now the Eco-complex Natural Conservation Area, has allowed some advances. But perhaps without these advances Tie Xi could not have become a serious candidate for the coveted UNEP Award and that the desire to win the award has been an important motivating factor, certainly recently, in sustaining them.

8(8) Ecological Counties

8(8) 1. Introduction

In 1994, seven government ministries (Ministry of Agriculture, N.E.P.A., the State Scientific and Technical Commission, Ministry of Finance, Ministry of Forestry, Ministry of Water Conservation and the Ministry of Planning) combined to establish a new CEA initiative at county level: fifty counties were chosen (at least one from each of the thirty provinces, autonomous regions or municipalities) to become exemplars of eco-agriculture. The counties, all of whom *applied* to be part of the project, were nominated at the provincial level and were chosen from a final list of 112. The primary responsibility was given to N.E.P.A. at national level to manage the initiative, Professor Bian Yousheng in Beijing took an overall organisational role and Professor Li Zhengfang in Nanjing was given direct responsibility for eleven counties to provide advice, technical support, training, supervision and inspection. Each county was given 20,000 yuan per year for three years to help them establish ecological agricultural projects and to select model villages.

According to Li Zhengfang, the scheme has three priorities: first, to ensure overall environmental planning modified according to local conditions, secondly to encourage integrated agricultural systems, i.e. the integration of agriculture with non-polluting industry, with animal husbandry, forestry, fishing and, if possible, biogas, and thirdly to improve the immediate rural environment by encouraging vegetation cover, increasing soil fertility, reducing soil erosion and reducing the use of chemicals. For Li Zhengfang and his team, the most important problem lies with the excessive use of chemical fertilisers and pesticides and hence the need to encourage their *appropriate* use, alongside organic fertiliser

Among the eleven counties chosen in the scheme for which Li Zhengfang was given responsibility were Dazu County in Sichuan Province and Simao County in Yunnan Province. I visited the EPBs of both of these counties in the summer of 1995.

8(8) 2. Dazu

8(8) 2.1 Introduction

Dazu is a county in southeast Sichuan Province (with a population of 116 millions, the most populous province in China), 150 kilometres west of Chongqing and linked to that huge metropolis on the Yangtse River by a new three-lane motorway built in 1995. The county has a population of a little less than a million (896,580 in 1994) and has agricultural land of 672,000 *mu*. There are approximately 220,000 households across 502 villages in 32 townships *(xiang)*, each household on average farming just three *mu*. Sichuan, with a GDP per head of 2516 yuan in 1994 (China Statistical Yearbook, 1996, p.45) is in the poorest third of provinces (twenty third of thirty). Dazu's income is average for neighbouring counties, approximately two-thirds of which comes from agriculture and one-third from industry. Given Sichuan's position in the centre of China, Dazu attracts no inward foreign investment, although it attracts increasingly large numbers of tourists from home and abroad to visit its exquisite Tang dynasty caves.

In the early 1980s, the HRS was introduced throughout the province - its governor was arch reformer and future Premier Zhao Ziyang - and was extremely popular amongst the farmers. The reforms were effective in increasing output as farmers sowed two or three crops rather than one. But agriculture in many parts of Dazu remained desperately poor, particularly in the hillier and more forested areas, where trees were cut down, despite the opposition of Dazu's leaders so to do.

8(8) 2.2 Implementation of ecological agriculture

Thus Dazu became interested in ecological agriculture in the early 1980s when the county leaders realized that, particularly in those hillier areas, both agriculture and the ecological environment were, according to Li Changke, in a "terrible state". There were no books nor propaganda about ecological agriculture at the time and experts from Chongqing including Ye Qianje were invited to act as consultants. Innovations occurred as a result of the ideas of the leaders of Dazu and were not initially popular amongst farmers. An initial experimental period of eco-agriculture began in 1984 involving thirteen villages and the initiative expanded in 1989 to include fifteen townships.

In each of the townships and villages involved a leader has been nominated who is specifically responsible for ecological agriculture, trained in the principles and practice of ecological agriculture who takes *commands* from the county environmental bureau and puts them into practice. The county provides education, technical guidance and support and there is a radio station in each village with newspapers and other propaganda to encourage ecological agricultural techniques. As incomes from the fields grew in the 80s, the villagers became increasingly willing to accept them. In 1994, Dazu was selected to become one of the fifty "ecological agricultural counties" to take part in the new nationwide initiative and was delighted with both the honour and the money.

Dazu was obligated to make an environmental plan to 1998. It designated three ecological areas, distinguished by altitude from sea-level, each characterized by a different and diversified basis to its economy: in the low-lying plains to the east, the major economic activities include rice growing, animal husbandry, fishing and processing industries, tobacco and biogas, in the mid-lying central belt of Dazu up to 400 m above sea level, activities include crop and meat processing, silkworm raising, iron handicrafts, especially knives, and tourism, based on the Tang dynasty caves, while in the hills above 400 m to the west, the economy is based on wheat, fruits and forestry, and the processing of wood products, including the handles for the knives, and bamboo products. In the 1994 plan, twelve eco-projects were established to cater for each element of the economy mentioned above.

In each geographical area, integration of planting, raising, and processing is encouraged and in accordance with ecological agricultural principles, more than ten different kinds of multi-level activities take place, including in the wet rice fields, combinations of rice, algae and fish or rice, mulberry trees and fish, in the dry fields, wheat, beans and potatoes, or wheat, peanut sand vegetables or wheat, tobacco and vegetables, in the forests, high evergreen trees with low, deciduous trees and in the ponds different fish at different levels. One of the more successful ecological agricultural initiatives has been the raising of fish in the wet rice fields. This has had three beneficial impacts: the fish have loosened the soil, they have eaten pests and their excrement is used on the fields. The success of this initiative has attracted many visitors and it has been extended to the whole province as a result.

Li Changke argues that the application of chemical fertilisers, which is approximately 10 per cent of the total fertiliser application, is lower in Dazu than neighbouring counties because of the earlier emphasis on ecological agriculture there. But since most households have theirs own pigs, grow green manure and raise their own silkworms there is no shortage of organic fertiliser. Throughout the county, there is rigid control of the use of pesticides: villages who want to use it must buy it off the county; all told, 450 m. tons of pesticides are used in Dazu every year. Forest cover has increased from 9.7 per cent immediately after the reforms in 1983 to 22 per cent in 1994, with plans to increase it further to 25 per cent in 1998.

With regard to biogas, there are 5,500 pits, mostly in the east, but all are at individual household level with no community level biogas digesters, which would be expensive to build and difficult to maintain. All have been built in the 90s and only a few villages have experimented with them, largely because the local availability of coal, wood and stalks persuades farmers they have no need of biogas. Li Changke suggests that a great deal more propaganda will be necessary for farmers to change their minds. Indeed, those pits that have been built have largely been the result of the same local 'commandism' which has responsible for the reduced use of pesticides in Dazu.

Given the low level of industrial development in rural Sichuan and the very high density of population, the land is cultivated intensively. Very few farmers are leaving the land and even where people are working in factories, they merely take time off during the busy farming periods. Although part of the income earned from industry is from coal mining, Dazu discourages other polluting industry and stopped the proposed siting of a paperworks in 1992 because of the pollution associated with its waste water. Tourism is being actively pursued and a hotel has recently been built by the county authorities.

According to Li Changke, many farmers are very pleased with rising standards of living based on working both in the fields and the factories, keeping pigs and raising silkworms. The silkworm raising is itself a special cooperative initiative of the county: 40,000 households have been encouraged to grow mulberry trees, up to 500 per *mu*. The leaves are taken back to the households to be eaten by silkworms, raised from the frozen larva supplied by the cooperative. The worms have a 50-day growth cycle, and each household operates the cycle three times a year. The cooperative provides the households with larva, buys up the worms and sells them to the silk factories. The worms' excrement may subsequently be collected by the household to feed its fish, to put into a biogas pit or to fertilise the fields.

In Dazu County, houses are now almost exclusively brick-built, mostly one-storey, and every village now has its own primary school. Li Chang Ke makes no particular boasts about Dazu and suggests its income levels are comparable with neighbouring counties: leaders from Liu Min Ying have twice come to Dazu and while Li Changke is impressed with the example of the latter, suggests that Liu Min Ying is in receipt of all sorts of financial and technical help not available to Dazu. But he remains optimistic about the prospects for ecological agriculture in his county.

8(8) 3. Simao

8(8) 3.1 Introduction

Simao is a border county in southern Yunnan province, a few kilometres to the north of Xishuangbanna. Yunnan is a long way from Beijing, in the far south west of China, bordering Burma to the west and Vietnam to the south. It is semi-tropical. The province has a population of 39 millions and although the Han are in the majority, Yunnan is home to many minority nationalities including the Bai, Dai, Yi and Naxi. It is a poor province, with an income per head of 2490 yuan in 1994 (China Statistical Yearbook, 1996, p.45) Only five provinces in China have lower incomes per head. Income came predominantly from agriculture (44.8 per cent) and forestry (14.8 per cent) in 1994.

Simao is high-lying: varying from 786 metres above sea level at its lowest point to 2,400 metres at its highest, Simao town, with a population of 80,000, lying at 1,301 metres above sea level. Simao County's total population in 1994 was 171,749, of which one quarter (36,946) were minorities, the majority of the latter living in the rural areas rather than in the towns. Although in a semi-tropical region, given the variety of altitude, Simao County enjoys "multi-level weather" and is heavily forested. Yunnan is itself an out-of-the-way province and although Simao town has had since 1988 a small airport which links the county to Kunming, Yunnan's capital city, it is still very cut off. The bus from Kunming takes two days and even from Jinghong, the capital of Xishuangbanna, a mere 150 kilometres away, six hours. Given the altitude and uneven terrain, there is no railway.

8(8) 3.2 Implementation of ecological agriculture

The Environmental Protection Bureau in Simao opened in 1988. It was keen to develop ecological agriculture from the start because the county's agriculture was in a poor and backward state, its primary concern being the extent of deforestation. According to Pu Maosen, the forest cover at Liberation was 80 per cent but there had been three periods of serious deforestation in Simao since then: in the late 1950s during the Great Leap Forward, in the late 1960s during the Cultural Revolution and the period immediately after the rural reforms in the early 1980s. Indeed forest cover in that latter period fell to 30 per cent as trees were chopped down and households grew crops instead, even on slopes, yet with unimpressive success and with mounting soil erosion. So in 1988, the county gave *orders* to every village collectively and thereby to every household individually to replant trees. Additionally, an agricultural science institute was established in Simao and the county provided farmers with a range of technical and scientific advice and support. Simao was made an official ecological agricultural county in 1994.

8(8) 3.3 Outcomes

The initial aim was to increase forest cover by 2 per cent a year. By 1995, it had got back up to a level of 45.2 per cent. And through the advice provided by the county farmers were persuaded to build terraces, to improve irrigation systems and to use less wood for cooking, replacing wood with either stalks or biogas. Altogether 1,500 household level biogas pits have been built since 1988 to provide power for cooking and lighting. It had originally been introduced in the early 1970s but there had been a lot of difficulties for farmers in providing sufficient inputs and in maintaining constant output and as a result it was stopped. And despite the county recently providing each household with 300 yuan as an incentive to build its own pit, biogas has spread slowly and it is difficult to persuade farmers to adopt it.

Ecological agriculture is encouraged through the development of the experimental sites and extension through training classes for farmers. In recent years, graduates from agricultural schools have been appointed to work alongside village leaders. In Simao, CEA is interpreted to mean the appropriate combination of traditional and new scientific methods to grow different things according to different, local conditions. The main crops grown in Simao have traditionally been tea, rubber, coffee, wet rice, maize and wheat. In the last few years, twenty eight 'green food' areas have been created by the EPB, all collectively *jiti*-owned and managed, growing a range of eco-foods, including tea and fruits such as strawberries and watermelons in the experimental areas.

Silkworm raising has recently been introduced, as has the raising of fish in paddy fields as well as in the 10,000 *mu* of fishponds. New seeds have been introduced and ecological agricultural practices, such as multi-layer planting/raising encouraged. Very little chemical fertiliser - less than five kilograms per *mu* - is used as a result of a variety of factors, not least of which being its initial cost plus the cost of its transportation in the mountainous areas. Also many of the least educated farmers are more ecologically minded and do not believe in chemical fertilisers. All farmers have animals of their own and have a ready supply of organic fertiliser as a consequence. But additionally in the mountainous areas there is a tradition of growing green manure, started during the Cultural Revolution but which has accelerated since the arrival of scientific and technical support in 1988.

Tea growing has always been an important activity in Simao and the tea-growing fields were not divided during the reforms of the early 1980s. As a result, villagers were given small plots of land *(kouliangtian)* to grow crops, fruits and vegetables for their own consumption and for the market, and were employed by the *jiti* to work in the fields producing tea. Nor were the fields of rubber trees divided, nor were the fields in mountainous areas. As a result, many households in Simao have very small farms of less than one *mu* per person, while earning a monthly wage from the *jiti*. It is difficult for the farmers to increase household output given the small size of their plots and the impossibility of using machinery and 'scientific' methods on their farms as a result. But tea and rubber production has not suffered and coffee, grown alongside rubber trees has expanded as Nestles has entered into a joint venture with Simao. Tea production is particularly vibrant; there are 45,000 *mu* of tea fields, producing 400 kilograms per *mu* and the tea is sold to other parts of China and abroad, including Australia.

Given Simao's remoteness from big centres of population, industry is growing only slowly, the main industrial activities including wood processing, leather and paper, the latter providing a pollution problem. According to Pu Maosen, the future of the environment is linked to the future of the economy: farmers will not be persuaded to engage in ecologically beneficial activities which aren't economically beneficial. So far, ecological agricultural initiatives have paid off and this perhaps explains why the *orders* to adopt CEA have not met with the sort of resistance other farmers have displayed when asked to do something they initially did not wish to do (cf. the experiences of local commandism in Guizhou where demands that farmers concentrate on cultivating tobacco even where that entails financial losses for them were, indeed, met with active resistance. The average income per head in 1994 was 3000 yuan, above the average for the area and for Yunnan generally and the county is already attracting visitors impressed with its progress to date. Simao is in the throes of big plans up to 1998 to increase the forest cover to 50 per cent, to increase total vegetation cover to 95 per cent, to increase the green food growing areas to 110,000 *mu,* to increase the number of households using biogas to 4,395. For Pu Maosen, the "future is ecological agriculture".

8(8) 4. Conclusion

The extension of ecological agriculture to counties was originally perceived as a three year experiment and according to Prof. Bian Yousheng and Wen Qiuhua, despite progress in some areas, the experiment has not been an unqualified success, given the size of counties in comparison to villages. According to Wen Qiuhua in 1996, who inspected several counties alongside Prof. Li Zhengfang on behalf of N.E.P.A., it is often difficult to persuade farmers to give up chemical fertilisers.

9 Conclusions

9.1 Introduction

From its reading of the claims made for CEA up to 1992, the World Bank suggested that the various trial sites for CEA:

> appear to be model rural communities in which the design of crop, livestock and fish production and the use of human and animal wastes and crop by-products result in a self-contained, sustainable system. Production of biogas as fuel and heavy or exclusive use of organic fertilisers appear to be main features. Proponents claim that yields and revenues on these sites are significantly higher than averages in other areas (World Bank, 1992b, p.87).

It went on to say,

> However it is not clear what inducements or incentives the sponsors provide to villagers to adopt these systems or whether these systems are financially attractive in the long run. Moreover, it is unclear how the findings and techniques are to be extended to the rural population at large. *It would clearly be worthwhile to further investigate this ambitious program and the conclusions emerging from it* (World Bank, 1992b, p.87, emphasis added).

My research involved the closer investigation of CEA sites called for by the World Bank in 1992 and a consideration of the appropriateness and ease of CEA extension into the wider Chinese countryside. The results are set out below.

9.2 UNEP 'Global 500' Nomination

Liu Min Ying (or to be more accurate, its leader, Zhang Zhenlin), He Heng, Xiao Zhang Zhuang and Teng Tou won the UNEP 'Global 500' award for environmental achievement in 1987, 1988, 1990 and 1993 respectively, while Zhang Jiashun, leader of Xiao Zhang Zhuang, additionally won the award on an individual basis in 1992. Between 1987, when the award was inaugurated, and 1993, there were sixteen Chinese winners and of these, no less than six were awarded to individuals or enterprises associated with CEA. Qian Wei and Tie Xi were provincial nominees for the award in 1995 and 1996 respectively. Dou Dian is a model Chinese village on many counts but has never been nominated for the UNEP 'Global 500' award.

The nominations are made first by the province to the National Environmental Protection Agency in Beijing and from there, if successful, to the UNEP. It is difficult to discover why one village rather than another has won the award: there can be little doubt that within the bureaucracy of NEPA, as in all Chinese bureaucracy, decisions are made with regard to political and administrative considerations, as a result of personal lobbying and in the interests of individual career advancement. Yet in my many interviews in the course of this research, including those at NEPA and at various more local environmental protection bureaux, only Prof. Cheng Xu in 1993, then of Beijing Agricultural University, intimated that there may have been more worthy winners of the award than Liu Min Ying. Even then, his criticisms only stretched to suggesting that Liu Min Ying had made too great an economic sacrifice and he did not question the environmental advances made.

9.3 Chinese Ecological Agriculture: is it Sustainable Development?

In the mid-1990s in each of the locations visited, CEA was operating at a different stage of its developmental cycle. Some villages (e.g. Liu Min Ying, He Heng) already had 10-15 years experience of CEA, for others (e.g. Qian Wei) it was still a relatively new initiative. Some (Liu Min Ying, Qian Wei, Tou Teng, Dou Dian) were operating a - broadly - archetypal ecological agricultural system which included a community-level biogas digester at its heart, others (Xiao Zhang Zhuang, He Heng) having used household level pits in the recent past had dropped biogas generation from the agenda. Tie Xi had never experimented with biogas at all. In some villages (Teng Tou, Qian Wei, Dou Dian) industry had a very high profile and was responsible for much the largest part of output, in others (Xiao Zhang Zhuang, He Heng, Tie Xi, even Liu Min Ying) agriculture, animal husbandry, forestry and fishing were still important income earners.

Thus, even in so small a sample there is a large disparity of experience of CEA. Nonetheless, to the extent that all displayed certain essential features associated with CEA, the following generalizations can be made:

(i) The greater availability of barnyard manure and/or biogas slurry and/or green manure allows increases in the organic content of the soil, improving long-run soil productivity.

(ii) The development of various multi-level and 'stereo' forms of cultivation improves the fertility of the soil while increasing output.

(iii) As a consequence of the reduced application of chemical fertilisers, agriculture becomes less 'petroleum based', reducing the dependence of agriculture on the exploitation of fossil fuels. Meanwhile the danger of chemical fertiliser as a long term polluter of the land and of water courses is reduced.

(iv) There is a reduction in the use of chemical pesticides and a consequent reduction in the potential damage to natural life associated with them.

(v) Increased forest cover resulting from planting more trees and cutting fewer of them down leads, locally, to improved soil quality, less soil erosion, more birds, more stable water tables and a greener, more aesthetically pleasing environment and, globally, to less risk of climatic disturbance.

(vi) Education as to the virtues of environmental care in the process of rural industrialisation, leads to the rejection of heavily polluting enterprises, ensuring cleaner air and water than would otherwise be the case.

(vii) Where applicable, the development of biogas and solar energy for household lighting and cooking reduces the demand for electricity and since this is generated in China primarily by burning low grade, high sulphur coal, it reduces, however slightly, the exploitation of fossil fuels and the consequent dangers of acid rain and global warming. Biogas slurry helps improve soil quality while cooking with biogas inside the household improves the health of villagers.

CEA, as discussed in Chapter 6, is not *organic* agriculture: the former is designed to integrate more fully into the local economic system than the latter and takes a more liberal attitude to the - sparing- use of chemical fertilisers. It does not operate according to the very restrictive standards laid down by IFOAM and will inevitably be rejected by some as a shabby and ultimately unhelpful compromise as a result. As Karin Janz suggests,

> (CEA) is not the same as IFOAM organic certified agriculture.... What China needs is a radical shift in its rural policy because the country has tremendous ecological problems (3/4/96, private correspondence).

However while organic agriculture may promise an ecologically sounder future than CEA, there is no serious possibility of its generalized introduction throughout China given present conditions. As Karin Janz herself recognises, 'China uses the highest amount of synthetic agrochemicals per area in the world' (3/4/96, private correspondence).Given this state-of-affairs, it is all but utopian to expect the adoption of organic agriculture in the Chinese countryside on a wide scale. In 1995, referring to the perceived need to build up the market for organic products before organic agriculture could take hold, Prof. Bian Yousheng jokingly referred to Prof. Li Zhengfang (then Director of the Organic Food Centre in Nanjing) as a mere 'salesman.' Meanwhile Zhou Shengkun, associate professor at CAID argued in 1996 that except in isolated localities organic agriculture did not fit current Chinese conditions and could not therefore be used as a generalized model for rural development. Organic agriculture may be environmentally purer than CEA, but if its generalized adoption in the present conditions of the Chinese countryside is unrealizable, if organic agriculture is simply economically unacceptable, then it can be discounted in a consideration of *sustainable development* across the generality of the Chinese countryside today.

CEA has been more widely adopted, however, and in the villages and counties studied has gone along with economic development which has provided higher levels of income and output, satisfied increasing consumer expectations, provided new forms of employment for those displaced from the land while having beneficial impacts on the natural environment, locally, regionally and globally. Chapter 2 discussed the concept of sustainable development and the possible contexts within which it could operate, noting Jacobs' advocacy of the importance of the 'environmental impact coefficient' falling faster than the rate of growth of GNP (Jacobs, 1991, p.42) for environmentally 'friendly' development to take place, and spelling out the three conditions under which this might be the case, viz. (i) a change in the composition of final output towards less environmentally damaging products (e.g. from goods to services), (ii) the substitution of less environmentally damaging factor inputs for more damaging ones (e.g. from fossil fuels to renewables) and (iii) an increase in the efficiency of resource use through technological progress (e.g. by increasing energy conservation efficiency). To the extent that CEA has achieved these three conditions in all the villages and counties visited, CEA would presumably get the thumbs-up from Jacobs and Ekins (1995,pp.25-29) and Lecomber (1975, p.42) as an environmentally acceptable form of economic development.

As a result, this thesis claims that CEA has the *potential* for sustainable development in the Chinese countryside. It is unsurprising that CEA has had environmentally beneficial impacts. To the extent that it has, it is clearly *technically* possible and *environmentally* friendly. What is important about this study, however, is the verification that the adoption of CEA *can* take place without any reduction in the pace of economic advance, indeed that CEA may accompany *unusually rapid* economic advance. If it can make money, it is adoptable and if adoptable, potentially sustainable in *social* terms.

9.4 Translating Potentiality into Actuality: the Initial Adoption of CEA

To the extent that the villages and counties studied varied considerably with regard to climate, altitude, soil type and terrain, it is clear that physical factors on their own are no barrier to the successful adoption of CEA, even though they will clearly affect the *type* of CEA chosen and climate will have an impact on the capacity to generate biogas at all. Political-economic factors are a different matter, however.

While each of the seven eco-villages adopted CEA at some stage in the 1980s or early 1990s, the prior social conditions for and manner of its adoption varied considerably from village to village. Indeed, as the case studies illustrate, there is substantial variation amongst them with regard to experience of environmental difficulties and recent political-economic history generally. Equally, as the case studies also highlight, there are differences with regard to the success of the adoption of CEA and the sustainability of particular practices, such as biogas generation, in them. Of course, physical geographical factors vary too. However, with regard to the political-economic preconditions for the successful adoption of CEA, the variety of experience described in the case studies, particularly in the crucial period between Liberation in 1949 and the 3rd Plenum in 1978, suggests that there are *none* which are absolutely necessary or sufficient. Nonetheless, there appears to be a range of factors which predispose a village (or indeed county) towards CEA adoption, which if present in sufficient numbers, is likely to make CEA not only more adoptable but more sustainable. The factors, in order of importance, are as follows:

9.4.1 A strong collective -jiti - organization

In the seven villages studied, agriculture was organized wholly on cooperative lines in four of them. At the onset of the reforms, two villages, Liu Min Ying and Dou Dian, refused to divide their fields and never did so, while two other villages, Qian Wei and Teng Tou, having divided their fields at reform collectivized them again soon afterwards. Thus, while family farming was reintroduced into China in the early 1980s to 98 per cent of the old communes and continues to operate across more than 90 per cent of the Chinese countryside today, farming is wholly collectivized in a *majority* of the villages studied. And in the three villages where agriculture is not, there are strong collective norms and influence. In Xiao Zhang Zhuang, the trees and waterways stayed collectively owned and managed and remain so while in Tie Xi the ecological forestry and ecological agriculture are collectively administered, even though households have their own plots for growing vegetables. In He Heng (as in Dazu and Simao) the reforms led to the chopping down of trees and resulting environmental damage, but the process provoked a rapid response from the leaders, resulting in the invitation to Prof. Li Zhengfang's team to repair the situation urgently and thus to the construction of a collective ecological agricultural farm and to collective reforestry. And today in He Heng, while the HRS continues to operate in part, the recent development of 'green' and organic food production has resulted from the land reorganisation in 1994 which brought a new 'green' farm under *jiti* control.

It is therefore clear from the experiences of the villages studied that, in Muldavins's words (1992, 1996a&b) the 'mining of ecological and communal capital' consequent upon the reforms which was apparent in many parts of the Chinese countryside (see Chapter 3) was either sufficiently well foreseen to do something about it beforehand, or if it did occur was retrieved by the leadership as a matter of urgency. The response was in every case a defence of or reversion to *collective* forms.

9.4.2 The question of leadership

When I asked officials at NEPA, in the Ministry of Education and in environmental protection institutes and bureaus what they considered to be the most important determining factor as to whether a village successfully adopted CEA, the most commonly held view was the importance of leadership, leadership which was not only imaginative and far-sighted, but firm and decisive. Once inside the villages, the same opinion was constantly iterated. And it was not just the adoption of CEA that involved strong and imaginative leadership. After all, in all the villages studied, it was quite clear that decisions were made top-down. While leaders made a point of remaining in touch with the feelings of villagers through the Party bureaucracy and, in some cases (e.g. Liu Min Ying and Teng Tou) through extensive public meetings, there were clear limits to democratic decision-making.

Research for this book did not involve a study of village leadership *per se*. Nonetheless such a study would clearly be of importance to an understanding of the political economy of Chinese villages generally, and of the feasibility of the adoption of such initiatives as CEA in particular. Amongst the seven villages studied there were many different kinds of leader. The term 'leader' - *lingdao* - is itself an ambiguous term. The village leader as such - *cunzhang* - sometimes overlaps with the position of Party secretary - *shuji* - sometimes not. In some cases, the prime mover, the *key* leader, has no official position in the village. This is the case in Xiao Zhang Zhuang where Zhang Jiashun rules the roost (is clearly 'the paramount' leader) yet has no village responsibility per se (he is presently deputy *shuji* of the county while his son is *shuji* of the township) and the *cunzhang* and *shuji* in Xiao Zhang Zhuang itself have subordinate roles. In Qian Wei, while the current *cunzhang* and *shuji* are more significant players, the former *shuji*, Xu Weiguo, now referred to as Xu *zhong* still holds paramount influence as President of the Board. In Dou Dian, He Heng, Xiao Zhang Zhuang and Liu Min Ying the leaders - *cunzhang*- are also currently (1997) the Party secretaries - *shuji* - although this has not always been the case even in the recent past and in Liu Min Ying, the formal leader, Zhang Zhanlin, is a figurehead and the real power and influence, certainly behind ecological agriculture has been deputy *shuji* and farm manager, Zhang Kuichang. In He Heng, from 1993 onwards neither *cunzhang* or *shuji* seems to have particularly important with regard to ecological agriculture allowing Zhang Tailin of the local EPB considerable space to dictate events. In Tie Xi, two brothers, Zhang Hongsheng and Zhang Hongtian, fill the roles of *shuji* and *cunzhang* respectively and apparently run a duarchy.

There are four key elements of the 'paramount' leader(s) identifiable as being important with regard to the adoption of CEA: (i) longevity, (ii) identification with past economic successes, (iii) proof of the ability to stand up to authority in the past and (iv) expressed *long-term* concern for the development of the village.

Re. (i) and (ii), in every village except He Heng, the current leader identified as being responsible for the introduction of CEA had been appointed before 1976. Two (Fu Jialiang in Teng Tou and Zhang Zhenliang in Dou Dian) were appointed before the Great Leap Forward and the rest assumed leadership either before or during the Cultural Revolution. In every case except one, therefore, the key leaders have been in office for twenty years or more, in two cases, for more than forty years. And the longevity of these leaders is clearly associated with, indeed explained by, their identification with past economic successes. The position of leader in a Chinese village is not directly comparable to his/her counterpart in the West. The leader is involved in a web of social relations, which according to Li Xiaoyun, Director of CIAD at China Agricultural University, is characteristic neither of capitalism nor socialism but 'orientalism'. The leader is formally a political position yet will perform an important 'social' role: he/she must be 'rooted in society and hence widely accepted, must have a clean reputation with regard to money and women and must have unchallenged intelligence' (1995, personal conversation).

(iii) If the above is the case, according to Li Xiaoyun, leaders will inspire enormous reserves of loyalty and will be able to take risks to achieve results. It is apparent that in the villages the subjects of this thesis, the key leaders maintain considerable social influence and inspire a degree of loyalty and respect (in some cases, love) more associated with a family head than a political leader. As a result, they are able to make crucial decisions (such as the adoption of CEA or the recollectivisation of the fields) in a more autocratic, final manner than would be the case if they did not. And leaders will take responsibility for the villagers in a manner more appropriate to this state-of-affairs. What is also significant about many of the leaders is that they were also associated with standing up to criticism (more often than not from ultra-leftist Party leaders at the county or district level for attempting rapid economic development through sideline activity and industrialisation well before the reforms of the 1980s). Zhang Zhenliang in Dou Dian, Fu

Jialiang in Teng Tou, Zhang Zhanlin in Liu Min Ying and the Zhang brothers in Tie Xi were all criticized during the Cultural Revolution for their leadership practices, the former being deposed for two years, yet they survived as a result of their popularity amongst the villagers. Fu Jialiang had to 'hide' the existence of a building materials works in Teng Tou during the Cultural Revolution in order to avoid any further censure to add to what he had earlier received from cadres at the county level for his exploits in growing fruit trees. Excepting He Heng, the leaders in all the villages researched for this thesis built up substantial reserves of loyalty many years ago, reserves which can be drawn upon whenever important decisions have had to be made since.

(iv) In that so many of the leaders have been around for so many years, it is unsurprising they should be concerned for the *long-run* prosperity of their villages. It was, however, a common response by the leader to my question as to why CEA was adopted in the first place, that *they* were taking a long view, while other leaders were concerned only for short-run prosperity. Indeed, It was frequently expressed that they were aware of the importance of the natural environment *before* environmental protection and CEA became popularized. This was specifically expressed by the leaders of four villages (Liu Min Ying, Xiao Zhang Zhuang, Teng Tou and Tie Xi).

As a bricoleur in a Chinese village, the importance of leadership graphically presented itself. It was not merely that officials or villagers themselves trumpeted the benign influence of their leaders: their very presence reinforced the message, a message identified through spoken and body language, grasp of essentials and evident competence. The leaders I met in all the eco-villages in this study were all *impressive* people, including Zhang Baocun, the leader of He Heng only from 1987 to 1993 (the only village not to have its key leader appointed before the end of the Cultural Revolution), who, no doubt because of his evident qualities, was quickly promoted to higher responsibilities in the township.

9.4.3 Money

All seven villages in this study were relatively rich *before* the adoption of CEA. That is not to say that many had not struggled (alongside most Chinese villages with poverty at some stage after Liberation) against poverty and had frequently understood the importance of a healthy ecological environment in combating that poverty. The latter is particularly true of Xiao Zhang Zhuang, He Heng and Teng Tou. But by the time of the *conscious* adoption of CEA, all had achieved a relatively high level of economic development. In the unique case of Qian Wei, where biotechnological development was causing potentially serious problems of pollution its adoption was a conscious decision to keep the village economically prosperous.

This factor is not unrelated to the two factors mentioned above: successful leaders of collective villages were, other things being equal, more likely to have sufficient funds to underwrite the short-run costs associated with the adoption of CEA, particularly the cost of establishing and running a community-level biogas digester, than their neighbours who weren't. Zhang Kuichang of Liu Min Ying, Xu Jianyu of Dou Dian and Guo Liangyao of Teng

Tou are all clear in their view that the existence of accumulated savings was one of the most critical factors separating the fortunes of their villages and neighbouring villages: without the internal financial resources to initially fund the developments they could not have been pursued. And shortage of funds today is the main reason why developments in their villages are not adopted elsewhere, they argue.

9.4.4 The presence of a 'change agent'

It seems clear that it is possible to adopt CEA without any help, or least with minimal help, from outside. It is the claim of Xiao Zhang Zhuang, Teng Tou and Tie Xi that this was originally the case, that CEA was adopted relying on the internal resources of the village alone. However, it seems just as clear that the successful adoption of CEA is much more likely *with* outside help.

Although the initial contact and relationship between the village and the 'change agent' was different in each case, it is clear that Liu Min Ying depended on the sponsorship of Prof. Bian Yousheng and the Beijing Research Institute of Environmental Protection for the initial construction of CEA, He Heng relied on the sponsorship of Prof. Li Zhengfang and his team at the Nanjing Institute of Environmental Science, Dou Dian on Prof. Cheng Xu at Beijing Agricultural University and Qian Wei on Shanghai's Fudan University. And *all* villages have obtained at least some outside support, if only from the county or provincial EPBs.

9.5 Translating Potentiality into Actuality: the Continuation of CEA

My research suggests that CEA continues to operate more comprehensively and successfully in some villages than others. Liu Min Ying, for example, was the first to adopt CEA in China yet it continues to prosper: it operates a system involving community level biogas and is the throes of *expanding* biogas generation; the application of chemical fertiliser continues to fall; industry and side-line activity thrive; the village is constructing new buildings year in, year out; eco-tourism is being actively developed. Dou Dian maintains two community-level biogas digesters and its commitment to CEA while continuing to make rapid economic development. Teng Tou, Qian Wei and Tie Xi also maintain very *explicit* commitment to CEA.

Xiao Zhang Zhuang and He Heng were also early adopters of CEA, but their experiences have been very different from, for example, Liu Min Ying and Dou Dian since then. On my visits in 1996, the leaders of *both* villages (without any prompting on my part) explained in a defensive fashion why biogas generation was not being pursued - largely on the grounds that economic development had obviated the need for biogas, given the alternative sources of power available today. Biogas could be encouraged only on *economic* grounds - thus it could be popularized only when there was no good alternative (e.g. as in the 1970s, before electrification). Meanwhile, both villages were continuing to apply chemical fertiliser (according to Bian Yousheng, in increasing amounts) and both seemed to be struggling to hold the line against the encroachment of polluting industry. Both still considered themselves models of CEA

- Xiao Zhang Zhuang erected a new statue signifying the village as a model environmental village in 1996 and He Heng has certainly regenerated itself as a producer of 'green food' - but the evidence is that CEA, *per se,* is taking a back seat when compared with other developments on the economic front.

My research suggests that the factors identified in 9.3 in predisposing villages to adopt CEA, i.e. collective agriculture, strong leadership, prior economic prosperity and the presence of a 'change agent,' also make its successful continuation and development more likely. I conclude from the evidence that the lack of (generalized) collective agriculture and animal husbandry in both He Heng and Xiao Zhang Zhuang, for example, is the primary reason for the lack of development of community-level biogas in the 90s, despite the enthusiasm both villages showed for it in the late 80s and early 90s, when they won the UNEP 'Global 500' award.

Although Xiao Zhang Zhuang continues to enjoy sustained, strong leadership, this is not obviously the case in He Heng and neither village benefited significantly from the patronage of powerful 'change agents' in the early 1990s. Both enjoy support from their local county and township environmental protection bureaux and He Heng has more recently enjoyed the help and guidance of the Nanjing Institute of Environmental Science once again with regard to the growing of 'green' products. But the evidence suggests that He Heng's weakening in its commitment to ecological agriculture in the early 1990s stemmed from the withdrawal of the Nanjing team from the village in 1989. Meanwhile, neither village enjoys the same sort of outside patronage available to Liu Min Ying from Prof. Bian Yousheng and to Dou Dian from the increasingly powerful Prof. Cheng Xu, in 1997 Head of the Education Department at the Ministry of Agriculture.

However, for CEA to be successfully developed once adopted, it is clear that it must be accompanied by and preferably associated with rapid economic development generally. Two further, interrelated factors encourage this development

9.5.1 Location

The political-economy of each village has been powerfully affected by its geographical location in relation to major centres of urban population. Liu Min Ying and Dou Dian are both within the municipality of Beijing while Qian Wei is inside the municipality of Shanghai and all are well integrated into the urban hinterland. Other villages are less well located: Teng Tou is several hours journey from Hangzhou, as is He Heng from Nanjing, but both are in relatively prosperous, densely populated regions. Xiao Zhang Zhuang and Tie Xi enjoy locations somewhat remote from large urban centres, however. I conclude from the evidence of my research that while a remote location does not appear to be a barrier to the *adoption* of CEA, it makes its continued development more difficult. The closer a village is to a municipal centre, the closer its likely links with powerful leaders and significant 'change agents', the more VIPS it is likely to attract, the more inward capital investment it is likely to receive, the closer it will be to rich markets for its products. The more remote the village, the less likely it will benefit from these factors, the slower its economic development is likely to be without respect to CEA and the less likely will CEA be associated with economic success. CEA stands no chance of adoption and continued development unless it promises that success. The differing locations of Liu Min Ying and Xiao Zhang Zhuang, for example, are clearly significant factors in explaining the differing levels of CEA development and indeed commitment to it in those villages since the adoption of CEA in the early 1980s.

9.5.2 *"the self-fulfilling prophesy"*

Those villages initially adopting CEA, particularly those which won prestigious awards for so doing, have attracted not only honour but attention from political leaders and potential investors. It is quite clear that one of the key motives for a village to adopt CEA in the first place is to receive the plaudits. The inward investment on the basis of a clean environmental record, however, must be considered a bonus. Liu Min Ying more than any other village has benefited in this way, although Dou Dian, Xiao Zhang Zhuang and Teng Tou have also done so. The more widespread CEA becomes, however, the less impact will this factor have.

9.6 The Translation of Potentiality into Actuality: some Conclusions

The evidence of this research is not optimistic for the extension of CEA in the present conditions of rural China. The conditions under which text-book CEA is likely to be adopted in a village and, once adopted, to thrive, are highly restrictive. In the first place, it is helpful that the village should be well led, relatively prosperous, enjoy outside assistance and be well located near a major urban centre. But additionally it is crucial that collective norms remain influential, preferably that agriculture and animal husbandry are run wholly collectively by the *jiti*. These conditions are extremely rare. I visited the environmental protection bureaux in both Daxing County (home of Liu Min Ying) and Chongming Dao (home of Qian Wei) each county having been nominated as one of fifty 'environmental counties' by NEPA in 1994. In each case, the officials (Wang Yuling, deputy director in Daxing County and Huang Yongzhou, project officer in Chongming Dao) suggested that while there was a range of CEA initiatives ongoing in their counties, only in Liu Min Ying and Qian Wei was there a complete CEA *system,* that this was because those villages operated as *jiti* and that without the *jiti,* a CEA *system* was out of the question.

9.7 Policy Implications

As explained in Chapter 6, the Chinese government has encouraged the promotion of ecological agriculture since 1982 and in 1996 claimed the existence of about 3,000 CEA sites of various kinds. Though this is a large number, it represents a tiny proportion of the Chinese countryside and suggests that despite the financial assistance from environmental institutes and agencies, the examples of model villages and government propaganda generally, CEA remains difficult to popularize and its adoption rare. And my research for this book suggests why, namely that the *conditions* for its successful adoption are restrictive and that in particular the decollectivization of the Chinese countryside has made the adoption of a CEA system an unlikely state-of-affairs. And therein lies a serious paradox. As suggested in Chapter 5, the break-up of the communes and the re - introduction of family farming is, in part, responsible for the present environmental unsustainability of Chinese agriculture, yet it also presents the greatest obstacle to the adoption of a more ecologically sound alternative.

In conventional economics literature in the west (and based on the pessimistic implications of the 'Tragedy of the Commons' (Hardin, 1980, p.100)), it is commonplace to argue for the extension of property rights as a means of combating environmental pollution and degradation. Hence the enthusiasm of Ross (1988, p.1) for market-based reforms, such as the granting of property rights, as a means of protecting the environment. However, as Redclift (1994, p.6) implies, such a policy is only likely to be of benefit to the environment if the individuals or institutions granted such rights (over the land, the forests, water resources or the sea) recognise a long-term benefit in protecting their property from pollution of degradation. This, however, is likely to be the case only in already economically 'developed' countries where environmental quality is normally given, particularly by the rich and powerful in those countries, a higher value than further material advances. In poor countries like China, however, the granting of property rights to -poor- individuals who see an immediate short-run benefit in polluting or degrading their property (e.g. by felling 'privatized' trees, exploiting 'privatized' minerals or washing washing heavy metals from 'privatized' electroplating workshops into 'privatized' or communal waterways) will have the opposite effect.

The importance of relative affluence to the perceived value of the environment is illustrated by the fact that it was the richest villages researched (e.g. Teng Tou, Qian Wei, Liu Min Ying, Dou Dian) that had developed CEA most fully. Not only are these all villages where the *jiti* runs agriculture, they are also rich enough villages to be able to *afford* environmental protection *in both senses*, viz. not only do they have sufficient communal savings to devote to the task but they are in a position to value environmental improvement sufficiently highly to *want* to do so.

In this light, it seems clear that the 'deepening' *(shenhua)* of the reforms so frequently called for by Chinese leaders in the 1990s must take these questions into account. While the 15th Peoples Congress in October 1997 argued for further wholesale privatization of state-owned industry, the further entrenching of property rights in most parts of the Chinese countryside will be at the expense of greater deterioration of the rural environment. Small, 'privatized' plots, whether *sirentian* or *kouliangtian,* are effectively too small to be efficient farms yet too big to be gardens or allotments. They remain unattractive economic propositions when employment in enterprises is available, while still involving villagers in considerable manual toil and time. The result of this is that farmers are less likely to have either the means *or the will* to engage in practices such as CEA.

9.8 The Land Question

There are therefore very real questions to be considered concerning the future of the landholding system if CEA is to be popularized. In one critical respect, there has already been a change: since 1994 the government has (although this has not been uniformly implemented across the country) encouraged the extension of household leases over the land for another 30 and in some cases 50 years. To the extent that farmers have a *long-term* interest in the land they till, they are that much more likely to use methods such as the application of organic fertilisers and inter- and multi-cropping, designed to maintain the fertility of the soil over time. But while this change has been welcomed by farmers and addresses one source of unsustainability, it has the negative effect of cementing the status-quo, of preserving small and increasingly uneconomic plots in aspic into the second quarter of the twenty first century.

Li Bingkun, an official of the State Council, is quoted under the headline "RURAL LAND SYSTEM NEEDS CHANGING" as saying,

> China has to encourage systems that facilitate scaled and efficient development of agriculture....the transfer of land rights is encouraged.....
> In actual practice, several new patterns have been developed. The most (popular) one is the so-called 'dual-land system' under which every farmer has his ration of land for food. The rest of the land, *connected and smooth,* is to be leased to people who are capable of scaled farming. In some ares, farmers (have) got the rights to dispose of the land they rent. They may, except for selling, lease the land again, join partnerships or shareholding collectives with it, and even lease it back to the land collective owner. (China Daily, 15/6/95, p.4, emphasis added).

Li Bingkun is clear that change should be based on the 'free will of farmers' but suggests that central and local government 'do their best to help perfect the systems' and adds,

> Right now, the key point is to allow transfer of the landuse rights so that all production resources can be distributed economically and efficiently. *And to put as much land back into scaled use, favourable conditions should be created for farmers so (that) many will give up their land thoroughly and willingly* (China Daily, 15/6/95, p.4, emphasis added).

This is not an isolated view. Shao Ning argues,

> A micro-economic mechanism for rural China must ..be able to solve the problems of agricultural operations *on an appropriate scale...* A central issue in China's agricultural transformation is the transfer of surplus labour, yet if labour is tranferred and *land is not concentrated,* labour transfer will not serve agriculture itself (1992, p.20).

9.9 Conclusion

An authoritarian return to collectivization is unthinkable and is clearly not an option, but the status quo for the next 30 or 50 years is not sustainable either. That status quo, where the rural economy, in the form of township-and-village enterprises, is frequently dependent on *collective* industry and services on the one hand yet on *privatized* agriculture on the other, creates an anomaly which offers no encouragement to sustainable agriculture in general or CEA in particular. When Mao exhorted the collectives to 'take grain as the key link, pay attention to animal husbandry, forestry, fish-raising and side-line occupations and develop an all-round economy' the latter was frequently forgotten in an orgy of grain monoculture. It is surely paradoxical that today, many villages and townships in the post-reform Chinese countryside are *collectively* going full steam ahead with sideline occupations and the all-round economy, whilst leaving the planting of grain - 'the key link'- to a *privatized,* precarious and increasingly unsustainable future, ruling out the extension of CEA and the promise of a more sustainable one.

Indeed, it is highly paradoxical that while, on the one hand, the Chinese leadership are calling for ever deeper reforms *(shenhua gaige),* with the implication of increasing marketization and expansion of private property rights, the same leaders have frequently chosen, for their models from which to learn in the 1980s and 1990s (in the way that Dazhai was chosen in the 1960s), villages which are wholly *jiti,* and where agriculture and animal husbandry are run collectively. In the mid 1980s, it was Dou Dian that became an exemplar, in the early 1990s (until the debacle in 1993 when its leader was tried for murder and sentenced to 20 years imprisonment, see Chen,1997, p.110)) Daqiu Zhuang in the suburbs of Tianjin, and currently (1997) it is Han Cun He, in Beijing's suburbs. While all three have owed their prosperity to different things (the latter to the formation of a huge building team cashing in on the current construction boom in Beijing), what unites them is collective management of agriculture. These kind of villages are examples of what are known as 'village conglomerates' (VCs). According to Chen (1997, p.109),

> The VC that has evolved from the former village brigade is a natural village-based communal collective with considerable diversification in agricultural production, industrial production and commercial activities. It is a collective because land, property and enterprises are owned by the VC and various activities are conducted under the leadership of the VC Party organisation and under the management of the village committee, VC membership is primarily limited to its natural village residents. Unlike state-run enterprises, the VC is responsible for its own success and for providing services and benefits for its own members.

According to Chen, the ten most wealthy villages nationwide (with income over 1b. yuan) in 1993 were *all* VCs (1997, p.109). In Shandong, 'almost all successful and fast-growing village economies ....are VCs' (Chen, 1998, p.73). Those who hail the land reforms as progenitors of rural development (and advocate their deepening to advance it further) seem either oblivious to this fact or to deliberately ignore it. Not all of these rich villages practise CEA, but all are in a position to do so, should their leaders advocate it.

It is instructive that Liu Min Ying, Qian Wei and Teng Tou, as well as Dou Dian, are also VCs, while He Heng, Tie Xi and even Xiao Zhang Zhuang are not a long way off. In each of the first four villages, economic development has occurred to the extent that each employs substantial numbers of workers from outside, in most cases with temporary *hukou* status. The employment of rural labour from outside is a common feature of industrialised villages in China but it is not the sole preserve of the fully privatised villages. In my researches it appears that the greater the income per head within the village, the greater the differential in payments to villagers and to those employed from outside (thus such a differential is high in Teng Tou and Qian Wei while being less in Liu Min Ying and Dou Dian).

In poor parts of rural China, such as those areas in which Aubert (1995) and Pennarz (1995) did their research, continued reinforcement of small-scale privatized agriculture may well *not* encourage CEA, although given the lack of attractive employment opportunities in enterprises, there may be no viable alternative to the status quo. But in the richer parts, particularly on the eastern seaboard, the faster more *collective* forms of land management can be encouraged, the better it will be both for agriculture in general and CEA in particular. There are likely to be no perfect models: in the 'multiple Chinas' (Muldavin 1996b, p.315) of today, different organizational forms are likely to suit different political-economic experiences. But if villages are to be both rich enough to *afford* the means of establishing an archetypal CEA system and to be *minded* to do so, agriculture (i) on a larger scale (ii) on a more collectivized basis and (iii) with a longer-term horizon is a *sine qua non.*

In the autumn of 1997, Xiong Xuegang, in an official Chinese journal, called for China to 'promote ecological agriculture' (1997, p.28). This will be very difficult to do unless Chinese leaders are prepared to address the structural, *political* changes that need to be made in terms of land management in the Chinese countryside that provide the necessary preconditions. That the process of land consolidation is itself dependent upon the precondition of the availability of non-farm employment in rural industrial enterprises means that the necessary changes are possible in the first instance only in the more industrially developed, richer, parts of China. Nonetheless, the central authorities must take the problem on board far more urgently and comprehensively than they have done so far and not merely leave it to the vagaries of local leadership.

References

Adams, W.M. (1990), *Green Development, Environment and Sustainability in the Third World,* Routledge, London and New York

Agriculture Investment and Research Group (1997), The Status Quo and Suggested Policies for Agricultural Protection in China, *Social Sciences in China,* Social Sciences in China Press, Beijing, Summer, Vol. 2, pp.88-97

Alcock, R. (1995), *Ecological Agriculture in China: Is it sustainable development?,* unpublished paper for MA Development Studies, University of Leeds

Alexandratos, N. (1995), *China's Projected Cereal Deficits in a World Context,* Paper delivered to ECARDC IV Conference, Manchester

Altieri, M. (1987), Agro-Ecology: The Scientific Basis for Alternative Agriculture, Westview Press, Boulder, IT Publications, London

Anon (1990), *Case Study on Eco-Farming in China with special emphasis on Rice,* paper given to me at CIAD, China Agricultural University

Anon (1993), *A brief introduction to Xiaozhangzhuang Eco-village in Yingshang County, Anhui Province, China,* given to me by Fuyang EPB in summer, 1993

Ash, R.F. (1991), The Peasant and the State, *China Quarterly,* No.127, pp.493-526

Ash, R.F. (1992), The Agricultural Sector in China: Performance and Policy Dilemmas during the 1990's, *China Quarterly,* No.131, pp.545-576

Aubert, C. (1995), *Grain and Meat Production in China; sustainability and change in two provinces, Henan and Jiangxi,* paper delivered to ECARDC IV Conference, Manchester

Bartelmus, P. (1986), *Environment and Development,* Allen & Unwin, London

Beckerman, W. (1974), *In Defence of Economic Growth,* Jonathan Cape, London

Bernstam, M. (1991), *The Wealth of Nations and the Environment,* Institute for Economic Affairs, London

Bernstein, B. (1979), unpublished MA(Educ.) lecture notes

Bian Yousheng (1988), *Liu Min Ying, Ecological Agricultural Construction,* (in Chinese), Beijing Publishing House

Blaikie, P. and Brookfield, H. (1987), *Land Degradation and Society,* Methuen, London and New York

Blaug, M. (1970), *An Introduction to the Economics of Education,* Allen Lane the Penguin Press, London

221

Booth, D. (1994), Rethinking Social Development, an overview, in *Rethinking Social Development, Theory, Research and Practice*, Booth D. [ed.], Longman, Scientific & Technical, Harlow

Boxer, B. (1991), China's Environment: Issues and Economic Implications, in *China's Economic Dilemmas in the 1990s: the problems of reforms, modernization and interdependence*, Study Papers submitted to the Joint Economic Committee, U.S. Congress, Washington D.C.

Brown, L. (1995), *Who Will Feed China? wake-up call for a small planet*, Earthscan Publications Ltd., London

Brown, P. (1996), *The Guardian*, 13/4/96, p.7

Brugler, B. & Reglar, S. (1994), *Politics, Economy and Society in Contemporary China*, Macmillan, London

Buck, J.L. (1964), *Land Utilization in China*, Paragon Book Reprint Corporation, New York

Bulmer, M. (1984), Concepts in the analysis of Quantitative Data, in M. Bulmer (ed.), *Sociological Research Methods*, 2nd Ed., Macmillan, London

van Buren [ed.] (1979), *A Chinese Biogas Manual*, ITDG Publications, London

Buttel, F. (1992), Environmentalism, *Rural Sociology*, Vol.57, No.1, pp.1-27

Cai, Y. and Smit, B. (1994), Sustainability in Agriculture: a General Review, in *Agriculture, Ecosystems and Environment*, No. 49, pp. 299-307

Carson, R. (1963), *Silent Spring*, Hamish Hamilton, London

Catton, C. (1992), *Tears of the Dragon, China's Environmental Crisis*, Broadcasting Support Services, Channel 4 Television, London

Catton, C. (1993), Great Leap Backward, *New Statesman and Society*, January 8, pp.28-30

Chambers, R. (1983), *Rural Development, Putting the Last First*, Longman, Harlow

Chen, W. (1997), Peasant Challenge in Post-Communist China, *Journal Of Contemporary China*, Vol.6, No.14, pp.101-116

Cheng Xu and Simpson, J. (1989), Biological recycle Farming in the Peoples Republic of China: The Dou Dian Village experiment, in *American Journal of Alternative Agriculture*, Vol.4, No. 1, pp.3-7

Cheng Xu, Han, C.R. and Taylor, D. (1992), Sustainable Development in China, in *World Development*, Vol.20, No.8, pp.1127-1144

Cheng Xu (1994a), *Selected Works of Professor Cheng Xu's Papers on Sustainable Agriculture*, Celebrating the Establishment of China (Beijing Agricultural University) Branch, World Sustainable Agricultural Association, Beijing

Cheng Xu (1994b), "Hard" and "Soft" Restraints for China to Sustain Agricultural Development and to follow the Conventional Modernization Approach of Agriculture and Deserved Alternative Way, in *Integrated Resource Management for Sustainable Agriculture,* CIAD, Beijing Agricultural University Press

Chesneaux, J. (1989), *La Modernite Monde,* La Decouverte, Paris

China (1988), *Constitution of the People's Republic of China,* Foreign Languages Press, Beijing

China Pictorial Publications [ed.] (1989), *Reform in Rural China,* China Pictorial Publishing Co., Beijing

China Statistical Yearbook (1992), State Statistical Bureau of the PR of China, Beijing

China Statistical Yearbook (1995), State Statistical Bureau of the PR of China, Beijing

China Statistical Yearbook (1996), State Statistical Bureau of the PR of China, Beijing

Chossudovsky, M. (1986), *Towards Capitalist Reconstruction, Chinese Socialism after Mao,* Macmillan, London

Conway, G. (1985), Agro-ecosystem Analysis, *Agricultural Administration,* 20, pp.31-55

Daly, H.E. (1980), Introduction to the Steady State Economy, in *Economics, Ecology Ethics,* (ed.) Daly, H.E., W.H. Freeman and Co., San Francisco

Dasmann, R.(1984), An Introduction to World Conservation, in Thibodeau, F. & Field, H. (ed.) *Sustainable Tomorrow,* University Press of New England, Hanover

Delman, J. (1990), Projecting Agriculture towards the Year 2000; Projecting the Unprojectable, in *Remaking Peasant China,* Delman J. et.al. [ed.], Aarhus University Press

Donaldson, P. (1973), *Economics of the Real World,* Penguin, London

Dobson, A. (1990), *Green Political Thought,* Unwin Hyman, London

Dou Dian (1994), *Doudian, The New Modernized Socialist Village,* The General Corporation of Agriculture, Animal Husbandry, Industry and Commerce, Fangshan, Beijing

E.I.U. (1996), *China, Mongolia,* Country Report, Economist Intelligence Unit, 1996, 4th quarter

Edmonds, R.L. (1994), China's Environment: Problems and Prospects, in *China, the Next Decades,* Dwyer, D. [ed.], Longman, UK

Edmonds, R.L. (1995), *Patterns of China's Lost Harmony,* Routledge, London & New York

223

Edmonds, R.L. (1996), The Three Gorges Dam: Recent Developments and Prospects, in *China Review*, No.4, Summer, pp.4-13

Ehrlich, P.R. and Ehrlich, Anne H. (1980), Humanity at the Crossroads, i n *Economics, Ecology, Ethics, Essays Toward a Steady-State Economy*, Daly, H.E. [ed.], W.H. Freeman & Co., San Fransisco

Ehrlich, P.R. and Ehrlich, Anne H. (1990), *The Population Explosion*, Hutchinson, London

Ekins, P. (1993), 'Limits to Growth' and 'sustainable development': grappling with ecological realities', in *Ecological Economics*, No.8, pp.269-288

Ekins, P. and Jacobs, M. (1995), Environmental Sustainability and the Growth of GDP: Conditions for Compatibility, in *The North, The South and the Environment*, Bhaskar V. and Glyn, A. [ed.], Earthscan Publications Ltd, London

Escobar, A. (1996), Post Structural Political Ecology, in *Liberation Ecologies, environment, development, social movements*, (ed.) Peet, R. and Watts, M., Routledge, London and New York

Esteva, G. (1992), Development, in *The Development Dictionary, A guide to Knowledge as Power*, Sachs, W. [ed.], Witwatersrand University Press, Johannesburg

Evans, Richard (1993), *Deng Xiao Ping and the Making of Modern China*, Penguin Books, London

Fang Ru-Kang (1995), *The Study of Eco-Farming Construction in China*, paper presented to ECARDC IV Conference, Manchester Business School

Fay, B. (1975), *Social Theory and Political Practice*, George Allen and Unwin, London

Foley, G. (1987), *The Energy Question*, Penguin, London, 3rd. Edition

Foley, G. (1991a), *Global Warming, who is taking the heat?*, Panos, London

Foley, G. (1991b), *Energy Assistance Revisited - a Discussion Paper*, Stockholm Environment Institute, Stockholm

Foley, G. (1992), Renewable Energy in Third World Development Assistance, Learning from Experience, in *Energy Policy*, April, pp. 355-364

Frank, A.G. (1966), The Development of Underdevelopment, *Monthly Review*, Sept. pp.17-30

Frank, A.G. (1992), 'Latin American Development theories revisited: a participant review', *Latin American Perspectives*, Issue 73, Vol.19, No.2, pp.134-145

Geertz, C. (1975), *The Interpretation of Cultures*, Hutchinson of London

Glaiser, B. (1987) [ed.] *Learning From China? Development and Environment in Third World Countries*, Allen & Unwin, London

Glaiser, B. (1990), The Environmental Impact of Economic Development, in *The Geography of Contemporary China, The impact of Deng Xiao Ping's decade*, Terry Cannon [ed.], Routledge, London & New York

Glaiser, B. (1995), *Environment, Development and Agriculture, Integrated policy though human ecology*, University College of London Press, London

Glaser, B. and Strauss, A. (1967), *The Discovery of Grounded Theory*, Altine, Chicago

Goldemberg, J. (1996), *Energy, Environment and Development*, Earthscan, London

Gorelick, S. (1977), Undermining Heirarchy: Problems of Schooling in Capitalist America, in *Monthly Review*, October, pp.20-33

Hall, D.O. and Rosillo, F. (1991), *Biomass Energy Resources and Policy*, Draft of a Contractor Report prepared for the Office of Technology Assessment, United States Congress, Washington D.C.

Han, C.R. (1989), Recent Changes in the Rural Environment in China, in *Journal of Applied Ecology*, No.26, pp.803-12

Hardin, G. (1980), The Tragedy of the Commons, reprinted in *Economics, Ecology, Ethics, Essays Toward a Steady-State Economy*, Daly, H. (ed.), Freeman, San Francisco, first printed in *Science*, Vol.162, pp.1234-1248, 13 December, 1968

Harding, H. (1987), *China's Second Revolution, Reform after Mao*, Brookings Institute, Washington D.C.

Hart, S. (1997), Strategies for a Sustainable World, in *Harvard Business Review*, Jan-Feb., pp.68-77

He Baochan (1991), *China on the Edge, the crisis of ecology and development*, China Books and Periodicals Ltd., San Francisco

Hecht, S.B. (1987), The Evolution of Agricultural Thought, in *Agro-ecology, The Scientific Basis of Alternative Agriculture*, M. Altieri, Westview Press, Boulder, IT Publications, London

Henwood, K. and Pidgeon, N. (1993), Qualitative Research and Psychology, in *Social Research, Philosophy, Politics and Practice*, (ed.) M. Hammersley, Sage, London and New Delhi

Hinton, W. (1990), *The Great Reversal, The Privatization of China, 1978-89*, Monthly Review Press, New York

Ho, S. (1994), *Rural China in Transition, Non-agricultural Development in Rural Jiangsu, 1978-1990*, Clarendon Press, Oxford

Hotelling, H. (1931), The Economics of Exhaustible Resources, *Journal of Political Economy*, No.39, pp.137-75

Howard, Pat (1988), *Breaking the Iron Rice Bowl*, M.E. Sharpe, New York

Huiting, R.,(1986), An economic scenario for a conserver economy, in Ekins, P. [ed.], *The Living Economy: a New Economics in the Making*, Routledge, London

Hutton, W. (1996), *The State We're In,* Vintage, London

IUCN (International Union for the Conservation of Nature) 1991, *Caring for the Earth,* IUCN/UNEP/WWF, Gland, Switzerland

Jacob,s M. (1991), *The Green Economy,* Pluto Press, London

Jahiel, A.R. (1997), The Contradictory Impact of Reform on Environmental Protection in China, *China Quarterly,* No.149, pp.81-103

Keleher, Daniel (1992), *Peasant Power in China, The Era of Rural Reform, 1979-1989,* Yale University Press, New Haven and London

King (1926), *Farmers of Forty Centuries,* Jonathon Cape, London

Koestler, A. (1967), *The Ghost in the Machine,* Picador, London

Kolakowski, L.(1993), An Overall View of Positivism, in *Social Research, Philosophy, Politics and Practice,* (ed.) M. Hammersley, Sage, London and New Delhi

Latouche, S. (1992), Standard of Living, in *The Development Dictionary, A guide to Knowledge as Power,* Witwatersrand University Press, Johannesburg

Lecomber, R. (1975), *Economic Growth versus the Environment,* Macmillan, London

Leeming, F. (1993), *The Changing Geography of China,* Blackwell, Oxford

Lele, S. (1991), Sustainable Development: A Critical Review, in *World Development,* V.ol.19, No.6, pp.607-621

Levi-Strauss, C. (1974), *The Savage Mind,* Weidenfeld and Nicholson, London

Li Bingkun (1995), Rural Land System needs Changing, in *China Daily,* 15/6/95

Li Kangmin (1991), Developing Ecological Agriculture in 21st Century China, in *Journal of Asian Farm Systems Association,* Vol.1, No.2

Li Lianfu (1994), The Status Quo and Prospects of the Devlopment of Green Food in China, in *Towards Organic Farming in China, Challenges for a Sustainable Development,* (ed.) Janz and Jingzhong), CIAD, Beijing Agricultural University

Li Wenjuan, (1995), *An Analysis of Sustainable Development in Crops Production,* paper delivered to ECARDC IV, Manchester Business School

Li Zhengfang (1994a), Ecological Agriculture in China, in Highlights *of the Summary Reports of Environmental Science and Technology Research in China,* Xiang Feng & Xiao Xing Ji [ed.], Nanjing Research Institute of Environmental Science, NEPA, Hohai University Press

Li Zhengfang (1994b), Organic Agriculture, a Global Perspective: The Challenges and Opportunities for China, in Towards *Organic Farming in China, Challenges for a Sustainable Development,* Proceedings of the First International Symposium on Organic Farming in China, Janz, K. & Ye, J. [ed.], CIAD, Beijing Agricultural University

Lin, J.Y, (1990), Institutional Reforms in Chinese Agriculture: Retrospect and Prospect, in *Economic Reform in China, Problems and Prospects*, Dorn, J. and Wang, X. [ed.], University of Chicago Press, Chicago

Lin, J.Y. (1994), Rural Reforms and Agricultural Growth in China, in *American Economic Review*, Vol.82, No.1, pp.34-46

Lin, J.Y., Huang, J.K. and Rozelle, S. (1996), China's Food Economy: Past Performance and Future Trends, in *China in the 21st Century: Long-Term Global Implications*, OECD, Paris

Liu Guoguang, Liang Wensen & Others (1987), *China's Economy in the Year 2000*, New World Press, Beijing

Lotspeich, R. and Chen Aimin (1997), Environmental Protection in the People's Republic of China, *Journal of Contemporary China*, Vol.6, No.4, pp.33-59

Luo, S.M. & Han, C.R. (1990), Ecological Agriculture in China, in Edwards, C.A. et.al. (ed.), *Sustainable Agricultural Systems*, Soil and Water Conservation Society, Ankeny

McMillan, J., Whalley, J., and Zhu, L.(1989), The Impact of China's Economic Reforms on Agricultural Productivity Growth, *Journal of Political Economy*, Vol.97, No.4, pp.781-807

Meadows, D., Meadows, D.,Randers, J., and Behrens, W. (1972), *The Limits to Growth*, Universe Books, New York

Meadows, D., Meadows, D., and Randers, J. (1992), *Beyond the Limits, Global Collapse or a Sustainable Future*, Earthscan Publications Ltd., London

Mishan, E. (1967), *The Costs of Economic Growth*, Penguin, London

Muldavin, J.S.S. (1992), *China's Decade of Rural Reforms: The Impact of Agrarian Change on Sustainable Development*, unpublished PhD thesis, University of California at Berkeley, USA

Muldavin, J.S.S. (1996a), Agrarian Reform in China, in *Liberation Ecologies*, Peet R. and Watts M. [ed.], Routledge, London and New York

Muldavin, J.S.S., (1996b), Impact of Reform on Environmental Sustainability in Rural China, *Journal of Contemporary Asia*, Vol.6, No.3, pp.289-319

Myrdal, J. (1965), *Report from a Chinese Village*, Heinemann, London

N.E.P.A. (1991), *China's Eco-Farming*, China Environmental Science Press, Beijing

N.E.P.A. (1992), *National Report of the PRC on Environment and Development*, submission to UN Conference on Environment and Development in Rio De Janiero, English version, Beijing

N.E.P.A./S.P.C (1994), *Environmental Action Plan of China, 1991-2000*, produced by the National Environmental Protection Agency and the State Planning Commission, China Environmental Science Press, Beijing

Nolan, P. (1988), *The Political Economy of Collective Farms*, Polity Press, Cambridge

Nolan, P. and Dong, F. (ed.) (1990), *The Chinese Economy and its Future*, Polity Press, Cambridge.

O'Connor, M. [ed.] (1994), *Is Capitalism Sustainable?*, Guildford Press, New York

Odum, H.E. (1971), *Environment, Power and Society*, John Wiley, New York

Oi, J. (1989), *State and Peasant in Contemporary China, the Political Economy of Village Government*, University of California Press, Berkeley, Los Angeles, Oxford

Ormerod, P.(1992), Waiting for Newton, in *New Statesman and Society*, August 28

Pan, J.H. (1992), *Economic Efficiency and Ecological Sustainability: A synthetic approach*, unpublished paper, University of Cambridge, Department of Land Economy

Party History Research Centre of the Central Committee of the Chinese Communist Party (1991), *History of the Chinese Communist Party*, Foreign Languages Press, Beijing

Peet, R. and Watts, M. (1996), Liberation Ecology, in *Liberation Ecologies, environment, development, social movements*, (ed.) Peet, R. and Watts, M., Routledge and New York

Peet, R. and Watts, M. [ed.] (1996), *Liberation Ecologies, environment, development, social movements*, Routledge, London & New York

Pennarz, J. (1995), *Adaptive Land-use Strategies of Sichuan Smallholders, Subsistence production and agricultural intensification in a land-scarce poverty area*, paper delivered to ECARDC IV Conference, Manchester

Per Ronnas (1994), *Economic Diversification and Growth in Rural China: the Anatomy of a 'Socialist' Success Story*, 1994 Reprint Series, No.118, Handelshogskolan 1 Stockholm, Reprinted from *The Journal of Communist Studies*, Vol.9, No.3, September 1993

Pezzey, J. (1992) *Sustainable Development Concepts, An economic analysis*, World Bank Environment Paper No.2, The World Bank, Washington D.C.

Pierce, D., Markandya, A. and Barbier, E.B. (1989), *Blueprint for a Green Economy*, Earthscan Publications Ltd., London

Pierce, D., Barbier, E.B. and Markandya, A. (1990), *Sustainable Development, Economics and Development in the Third World*, Earthscan Publications Ltd., London

Pierce, D. [ed.] (1991) *Blueprint 2*, Earthscan Publications Ltd., London

Porritt J. (1984), *Seeing Green*, Blackwell, Oxford

Potter, S. and Potter, J. (1990), *Chinese Peasants: anatomy of a Revolution*, Cambridge University Press, England

Putterman, L. (1993), *Continuity and Change in China's Rural Development*, Oxford University Press, Oxford

228

Qu Geping (1991a), *Environmental Management in China*, UNEP, China Environmental Science Press, Beijing

Qu Geping (1991b), *The Review and Prospect of Eco-Farming Construction in China*, China Environmental Science Press, Beijing

Redclift, M. (1987), *Sustainable Development, Exploring the Contradictions*, Routledge, London and New York

Redclift, M. (1990), Beyond the Buzzword: Defining Sustainable Development, in *New Ground*, Spring, pp.4-5

Redclift, M. (1994), Reflections on the 'sustainable development' Debate, *International Journal of Sustainable Development and World Ecology*, Vol.1, pp.3-21

Redclift, M. and Sage, C. (1994) (ed.), *Strategies for Sustainable Development: Local Agendas for the Southern Hemisphere*, John Wiley & Sons, London

Riskin, Carl (1987), *China's Political Economy, the quest for development since 1949*, Oxford University Press, New York

Robinson, W. (1969), The Logical Structure of Analytic Induction, in *Issues in Participant Observation*, [ed.] McCall, G. & Simmons, J., Addison-Wesley, London, Reading, Massachusetts

Ross, L. (1988), *Environmental Policy in China*, Indiana University Press, Bloomington.

Ross, L. and Silk, M. (1987), *Environmental Law and Policy in the Peoples Republic of China*, Quorum Books, New York, Westport, Connecticut, London

Rozelle, S. (1996), Stagnation without Equity: Patterns of Growth and Inequality in China's Rural Economy, *The China Journal*, No.35, pp.64-92

Sachs, W. [ed.] (1992), *The Development Dictionary, A Guide to Knowledge as Power*, Witwatersrand University Press, Johannesburg

Schumacher, E.F. (1973), *Small is Beautiful, A Study of Economics as if People Mattered*, Abacus, London

Seabrook, J. (1994), *Victims of Development, Resistance and Alternatives*, Verso, London and New York

Shao, N. (1992), Development and Reform: China's Agriculture in the 1990's, i n *Social Sciences in China*, Vol.XIII, No.2, pp.16-22

Sinkule, B.J. and Ortolano, L. (1995), *Implementing Environmental Policy in China*, Praeger, New York

Smil, V.(1984), *The Bad Earth*, M.E. Sharpe, New York

Smil, V. (1987), Land Degradation in China: an ancient problem getting worse, in *Land Degradation and Society*, Blaikie, P.and Brookfield, H. [ed.], Methuen, London and New York

229

Smil, V. (1992), China's Environment in the 1980s: some critical changes, in *Ambio*, Vol.21, No.6., pp.431-436

Smil, V. (1993), *China's Environmental Crisis, An Inquiry into the Limits of National Development*, M.E. Sharpe, New York and London

Smith, R. (1997), Creative destruction: Capitalist Development and China's Environment, *New Left Review*, No.222, pp.3-40

Snow, Lois Wheeler (1981), *Edgar Snow's China*, Random House, London

State Council (1994a), *China's Agenda 21 - White Paper on China's Population, Environment and Development in the 21st Century*, China Environmental Science Press, Beijing

State Council (1994b), *Introduction to China's Agenda 21*, China Environmental Science Press, Beijing

State Council (1996), *Environmental Protection in China*, Information Office of the State Council of the PR China, Beijing

Statistical Survey of China (1997), in Chinese, State Statistical Bureau of China, Beijing

S.P.C/S.S.T.C (1994), *Priority Programme for China's Agenda 21, 1st Tranche*, State Planning Commission, State Science and Technology Commission, Beijing

Sutcliffe, R.(1995), Development after Ecology, in *The North, the South and the Environment*, Bhaskar and Glyn (ed.), Earthscan Publications, Ltd., London

Tao Siming (1993), *The Ecological Agriculture and its Development in China*, unpublished paper given to me by the author at NEPA, Beijing.

Therborn, G.(1977), *Science, Class and Society*, New Left Books, London

Tian Xueyuan (1997), Sustainable Development of the Population, the Economy and the Environment, in *Social Sciences in China*, Autumn, pp.29-35

Tisdell, C. (1988), Sustainable Development: Differing Perspectives of Ecologists and Economists, and Relevance to LDCs, *World Development*, Vol.16, No.3, pp.373-384

Turner, R.K. (1991), Environmental Economics, in *Developments in Economics*, Causeway Press, Ormskirk

Unger, J. (1985), The Decollectivisation of the Chinese Countryside, in *Pacific Affairs*, pp.585-606

Unger, J. (1988), State and Peasants in Post-Revolution China, in *Journal of Peasant Studies*, pp.114-135

UNDP (1990), *Human Development Report*, Oxford University Press, Oxford & New York

UNDP (1996), *Human Development Report*, Oxford University Press, Oxford & New York

UNEP (1995), *United Nations Environmental Programme, Global 500 winners*, posted to me from the UNEPs headquarters in Niarobi, December 22, 1995

van Buren, A. (ed.) (1979), *A Chinese Biogas Manual*, ITDG Publications, London

Vermeer, E.B. (1990), Management of Environmental Pollution in China: Problems and Abatement Problems, in *China Information*, Volume V, No.1, pp.34-65

Vermeer, E.B. (1995), An Inventory of Losses Due to Environmental Pollution: Problems in the Sustainability of China's Economic Growth, in *China Information*, Vol.X, No.1, pp.19-50

Wang Guichen, Zhou Qiren et.al. (1985), *Smashing the Communal Pot*, New World Press, Beijing

Wang Limin And Davies, J. (1996), *Can China feed its people into the Next Millenium: Projections for China's Grain Supply and Demand to 2010*, Paper delivered to 8th Annual Conference of Chinese Economic Association, London School of Economics

WCED (1987), *Our Common Future*, Oxford University Press, Oxford and New York

Wehrfritz, G. (1995), Grain Drain, *Newsweek*, May 15

White, G. (1993), *Riding the Tiger, The Politics of Economic Reform in Post-Mao China*, Macmillan, London

World Bank (1991), *The Challenge of Development*, World Bank Development Report, Oxford University Press

World Bank (1992a), *Development and the Environment*, World Bank Development Report, Oxford University Press

World Bank (1992b), *China Environmental Strategy Paper*, for official use only, Document of the World Bank, Washington D.C.

World Bank (1992c), *China, Strategies for Reducing Poverty in the 1990s*, World Bank Country Study, Washington D.C.

World Bank (1995), *Poverty in China, What do the numbers say?* Background Note, Washington D.C.

World Bank (1996), *From Plan to Market*, World Bank Development Report, Oxford University Press

von Wright, G.H. (1993), Two Traditions, in *Social Research, Philosophy, Politics and Practice*, (ed.) M. Hammersley, Sage, London and New Delhi

Xiao Zhang Zhuang (1993), *A Brief Introduction to Xiaozhangzhuang Eco-Village in Yingshang County*, (in English) publisher and author unnamed

Xiong Xuegang (1997), Thoughts on Developing China's Agriculture into an Industry, in *Social Sciences in China*, Autumn, pp.21-28

Yao Jianfu (1995), *Structures of Rural Production, Agricultural Technology and Rural Labors in China*, Paper delivered to ECARDC IV Conference, Manchester Business School

Yao Shujie (1994), *Agricultural Reforms and Grain Production in China*, Macmillan, London.

Zhang Kuichang (1994), *Glorious Twenty-five Years, Glory in the Future*, speech to journalists, September, translated into English by Xiong Ying

Zhou, Q.R (1996a), *The Rural Reform on Land Tenure in China*, Working paper on Farm Privatization in Uzbekistan for the World Bank in Tashkent, May 1996

Zhou, Q.R.(1996b), *Agricultural Reform: Property Rights and New Organization*, Paper delivered at the Annual Conference of the Chinese Economics Association, December 1996, at the London School of Economics

Zweig, D. (1989), *Agrarian Radicalism in China*, Harvard University Press, Cambridge, Mass. & London

Printed and bound by CPI Group (UK) Ltd, Croydon, CR0 4YY

22/10/2024

01777623-0004